TOPICS IN PLASMA
DIAGNOSTICS

TOPICS IN PLASMA DIAGNOSTICS

I. M. Podgornyi

Institute of Cosmic Studies
Moscow, USSR

Translated from Russian

PLENUM PRESS • NEW YORK–LONDON • 1971

Igor' Maksimovich Podgornyi was born in 1925 and was graduated from Kharkov State University in 1950. From 1951 to 1957 he worked in the I. V. Kurchatov Institute of Atomic Energy and in 1958 was awarded the Lenin Prize for his work on high-power-pulsed discharges. He lectured at Moscow State University from 1954 to 1963 and since 1967 has worked at the Institute of Cosmic Studies.

The original Russian text, published by Atomizdat in Moscow in 1968, has been corrected by the author for the present edition. The English translation is published under an agreement with Mezhdunarodnaya Kniga, the Soviet book export agency.

Игорь Максимович Подгорный
ЛЕКЦИИ ПО ДИАГНОСТИКЕ ПЛАЗМЫ
LEKTSII PO DIAGNOSTIKE PLAZMY

Library of Congress Catalog Card Number 72-137010

ISBN-13: 978-1-4684-0726-6   e-ISBN-13: 978-1-4684-0724-2
DOI: 10.1007/978-1-4684-0724-2

© 1971 Plenum Press, New York
Softcover reprint of the hardcover 1st edition 1971

A Division of Plenum Publishing Corporation
227 West 17th Street, New York, N. Y. 10011

United Kingdom edition published by Plenum Press, London
A Division of Plenum Publishing Company, Ltd.
Donington House, 30 Norfolk Street, London W.C. 2, England

Translator's Preface

The present volume is essentially a qualitative survey of modern trends in the diagnostics of high-temperature plasmas, with particular orientation toward laboratory plasmas of interest in connection with research in controlled thermonuclear fusion. Among the broad topics considered are probe diagnostics, optical methods (including the use of lasers and holography), microwave diagnostics, and diagnostics with particle beams. Having information on these methods available in compact form and in one place, as is the case in the present volume, should make it possible to evaluate different diagnostic approaches to specific problems. The volume will be useful as an introduction for advanced students making their first contact with experimental plasma physics and for physicists and engineers who are entering the field and desire a rapid survey of principles and modern trends in the diagnostics of high-temperature plasmas.

Foreword to the American Edition

The material in this book is based on lectures given at Moscow State University. It is intended to acquaint the reader with the basic aspects of plasma diagnostics and contains information required for the experimental physicist who wishes to carry out straightforward measurements of laboratory plasmas. It will be evident that in choosing the material we have been guided primarily by the scientific interests of the author, and the great bulk of the material is based on work carried out in the USSR.

During the time in which this book was in press in the Soviet Union, a similar work appeared in the United States, edited by Huddleston and Leonard. Their book reflects primarily the work of American laboratories, and its papers are intended as a survey of a number of methods of investigating high-temperature plasma.

These books are not duplicates, but, rather, supplements of one another.

It is the hope of the author that this translation will be of interest to the American reader and will provide a useful addition to the literature on plasma diagnostics in the USA.

Foreword

The beginning of experimental research in the field of high-temperature plasma physics may be taken to be the year 1951, when physicists from the Soviet Union and the USA, independently of each other and on a wide scale, started work on the heating and confinement of plasma. These investigations were started in connection with the problem of controlled thermonuclear fusion and, at that time, were carried out under conditions of strict secrecy. In 1956 the security barrier was broken in a report by I. V. Kurchatov in England, this report containing results obtained in plasma research in the laboratory of L. A. Artsimovich.

Even the first experiments indicated that the behavior of a plasma confined by a magnetic field is anything but that which had been predicted theoretically. In effect, the plasma was a patient suffering from an unknown disease, while the role of the experimenter was more or less that of a doctor who did not have at his disposal any other diagnostic instrument than a thermometer. Under these conditions, obviously, it was not possible to make anything like a proper diagnosis. In order to discover the reasons for the rapid loss of plasma across the magnetic field, first of all it was necessary to develop and apply a large arsenal of diagnostic instruments, which would more or less determine uniquely the state of a medium at temperatures of millions of degrees. Having this information available one might then hope to make a comparison between the experimentally observed effects and the theoretical predictions.

Many methods of investigating high-temperature plasma, which became a prime requisite in all work on plasma physics, have been borrowed from related fields of physics. This is certainly the case for astrophysics (the determination of temperature and density from the analysis of the shapes of spectral lines), classical discharges (Langmuir probe), and, finally, electronics (the measurement of the electrical parameters in powerful pulsed discharges).

Methods borrowed from other fields of physics were soon found to be inadequate and it became necessary to develop methods that were designed specifically for measuring the parameters of laboratory plasma. Thus, the diagnostics of high-temperature plasmas developed into a separate, rapidly developing field of applied physics, making use of all present-day theoretical possibilities. In turn, progress in the field of plasma diagnostics has had an effect on other fields of physics. For example, plasma spectroscopy has been advanced significantly, and it is now difficult to say which is the bigger stimulation to the development of this field, laboratory plasma research or astrophysical observations. New methods of diagnostics have appeared in connection with work on controlled thermonuclear reactions, these being based on the use of electromagnetic radiation in various wavelength regions. There has also been a concomitant development of plasma research methods that make use of charged-particle beams and neutral-particle beams. Finally, in recent years, with the appearance of lasers and masers, plasma physics has obtained a new diagnostic tool with which it is possible to obtain independent determinations of density and electron temperature. In addition, great value is still attached to photographic methods involving high-speed photography; these provided the first discoveries of plasma instability. The use of electron-optical image converters has also made it possible to increase the rate of picture taking by an order of magnitude.

The question of the relative importance of any given method of measurement in plasma diagnostics always leads to involved discussions and obviously can never be resolved uniquely. The answer to this question depends on the time and is obviously subjective to a large degree. It would be improper to say that there is a strict proportionality between the length of various sections in this volume and the value of the methods discussed in them. In the present work we have devoted relatively little space to the question of microwave measurement of plasma parameters since the Soviet literature contains recent books in this field of plasma diagnostics (V. D. Rusanov, Modern Methods of Plasma Research, Gosatomizdat, Moscow, 1962, and A. V. Chernetskii, O. A. Zinov'ev, and O. V. Kozlov, Methods and Instruments for Plasma Research, Gosatomizdat, Moscow, 1965).

Within the space limitations of the present survey it is impossible to consider in detail the various methods of plasma diagnostics that are presently being used. Many of these methods are rather restricted in scope and are of interest only to a rather limited number of readers; these readers can obtain information on such methods in papers that appear in the periodical literature.

This book represents a development of lectures presented by the author to students in the Physics Department at Moscow State University who specialize in the physics of high-temperature plasmas. The material has been chosen so as to be useful not only to students who are studying plasma physics, but to experimental plasma physicists in general. For this reason we have included a certain amount of handbook material which is needed for analysis of the experimental measurements. Many of the formulas are given with numerical coefficients in order to facilitate their use.

The author is indebted to L. A. Artsimovich, E. K. Zavoisky, Yu. K. Zemtsov, G. G. Managadze, G. V. Sholin, and V. P. Smirnov for discussions of various parts of the book.

Contents

Chapter 1

Oscilloscope Measurements of Current and Voltage in Pulsed Thermonuclear Devices

Among the various methods for confinement and heating of a plasma a significant place is occupied by those which make use of the passage of intense current pulses through the plasma. In typical thermonuclear devices the discharge current can reach millions of amperes and the induced voltages can be of the order of tens of kilovolts. It will be obvious that the standard methods used in everyday electronics are not suitable for measurements of such large currents. The problems involved in the measurement of such currents and voltages are further complicated by the fact that the pulse length is generally very short. For example, in plasma heating by fast single-stage compression in a pulsed discharge the pulse length is determined by the inertial confinement time, generally being of the order of a microsecond. It will be evident that if one wishes to obtain information on the development of a process that occurs in a thermonuclear reactor it is not sufficient merely to know the general characteristics of the pulse, such as its amplitude and length; rather, one must have a complete picture of the time variation of the current. In other words, the measurement must provide an oscilloscope trace of the current pulse with well-defined scales for both coordinates (current and time). Furthermore, the method that is used must be such that the measurement has little or no effect on the processes that occur in the discharge chamber itself. In addition, it is desirable that the measurement circuits have no direct metal contact with the discharge circuit. The elimination of such contact reduces the danger of breakdown in the measurement apparatus and also reduce the level of noise that arises because of parasitic coupling.

All of the requirements listed above can be met by the Rogowsky loop, used in conjunction with an integrating circuit and a pulsed oscilloscope. The essence of this method of measurement is the following. Around a conductor carrying a current is located a toroidal coil which is terminated in a small resistance R, the voltage drop across the resistance being recorded by an oscilloscope. A change in the current I_1 flowing in the conductor that threads the torus induces an electromotive force in the toroidal solenoid. The current in the solenoid I_2 is determined from the following relation:

$$I_2 R + \frac{1}{c^2} L_{22} \frac{dI_2}{dt} + \frac{1}{c^2} L_{21} \frac{dI_1}{dt} = 0, \qquad (1.1)$$

where L_{22} is the inductance of the toroidal solenoid and L_{21} is the coefficient of mutual inductance between the toroidal solenoid and the circuit carrying the current I_1. Solving this equation under the assumption that the voltage drop in the ohmic resistance is much smaller than that across the inductance, we can easily show that the voltage drop measured across the resistance R is determined by the number of turns of the solenoid, being independent of geometric factors,

$$V = \frac{I_1 R}{n} . \qquad (1.2)$$

In the derivation of this relation it is assumed that the ohmic resistance of the solenoid is small compared with the measurement resistance R.

The voltage drop V is independent of the orientation of the Rogowsky loop with respect to the current-carrying conductor. The voltage drop vanishes if the total current through the toroid vanishes regardless of whether the loop is close to the current-carrying conductor or not. This feature indicates the relative immunity of the system to electromagnetic noise.

The noise immunity of the Rogowsky loop can be made even greater if it is made in such a way as to be insensitive to variations

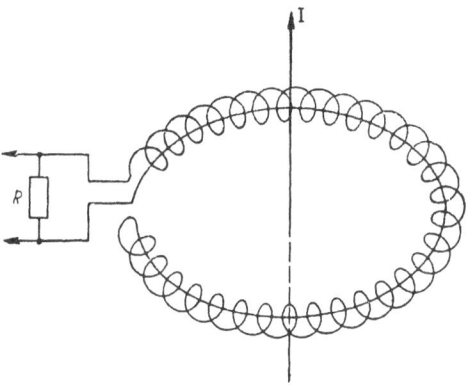

Fig. 1.1. Measurement of current with a Rogowsky loop.

in magnetic field for which the lines of force intersect the plane of
the torus. A magnetic field along the axis of a discharge current
is frequently used for stabilization purposes. In addition, such a
field can arise by virtue of hydrodynamic instabilities in a plasma
pinch. The sensitivity of the Rogowsky loop to a field of this kind
can be reduced to a minimum by means of a "reverse turn" located
inside the solenoid. The construction of a Rogowsky loop with a
reverse turn (Fig. 1.1) is also convenient in that it generally facil-
itates the mounting of the device in a system. The coupling between the Rogowsky loop and the oscilloscope is generally by
means of a coaxial cable, in which all connections are made by
coaxial elements. In order to suppress oscillations arising from
reflections at the end of the cable it is desirable to terminate the
cable in a matched load.

Special attention is given to careful shielding of the measurement resistance R. It is usually desirable to connect the solenoid
to the cable by a small coaxial feeder which contains the resistance; the value of the resistance is usually several tenths of
an ohm. The value of this resistance must be smaller than the
inductive reactance of the solenoid; on the other hand, it must be
large enough so that the signal fed to the oscilloscope can be detected without a preamplifier. If this requirement is not met additional difficulties arise that are associated with the increase in the
level of electromagnetic noise due to the presence of the preamplifier.

The noise immunity requirement is not the only one that must be kept in mind in the construction of the Rogowsky loop. If the curve that is traced out by the oscilloscope beam is to correspond to the dependence of the current on time the resistance of the loop must be negligibly small compared with its inductive reactance. Specifically, this requirement is assumed in the deviation of Eq. (1.2). Unfortunately, the condition $R \ll \omega L$ cannot always be satisfied for all values of ω (obviously here we are considering loops whose dimensions are such as to be convenient for use in practical devices). This condition is easily satisfied for a loop of reasonable dimensions at frequencies that exceed 10^3-10^4 Hz. The inductive reactance can also be increased by introducing a ferrite core in the solenoid. Rogowsky loops with ferrite cores have been used successfully at frequencies below one hundred Hertz. In the low-frequency region, and with loops of small dimensions, it is found that the integration of the signal from the solenoid can be realized through the use of an RC circuit whose time constant is large compared with the time constant of the process being investigated. As a rule, an integrator of this kind implies the use of an additional amplifier. The Rogowsky loop with the RC integrating circuit is frequently used in research on pulsed discharges stabilized by a magnetic field.

The high noise immunity, the reliability of the results, and the simplicity of construction make the Rogowsky loop one of the most popular instruments in thermonuclear research.

In order to investigate the behavior of a plasma column which is detached from the walls, in addition to measuring the current curve, it is generally necessary to know the time dependence of the potential drop in the plasma. A curve of the voltage can easily be obtained by use of an ohmic divider whose output goes to a pulsed oscilloscope (for example, the OK-17M). The signal from the Rogowsky loop is applied to the deflection plates of one of the oscilloscope beams while the signal from the divider is applied to the other plates. As a result, a single photograph can provide oscilloscope traces of the current and voltage with the proper phase relations. In making exact measurements, the accuracy of the relative phases can be monitored by means of time markers which are applied to both beams of the oscilloscope.

The best oscilloscope traces of the voltage are obtained if use is made of a matched load at the cable; in this case, the signal from the divider is taken from a resistance whose magnitude is equal to the characteristic impedance of the cable. Thus, the output resistance of the divider is known and its total resistance is determined by the amplitude of the voltage being measured and the sensitivity of the oscilloscope. When measuring a voltage, or a current, it is desirable to avoid the use of amplifiers and to make all cable connections directly to the deflection plates of the oscilloscope.

The ohmic divider is usually made from conventional resistances used in electronics; when operated under pulsed conditions these allow large power overloads. The loop voltage in circular discharges can be measured by means of a wire loop which is wound upon the circuit being measured. Depending on the magnitude of the loop voltage the ends of the loop are connected directly to an oscilloscope or through an ohmic divider.

We now consider a typical analysis of oscilloscope traces of current and voltage. Making use of synchronized oscilloscope traces, by graphical integration we can compute the energy introduced into the discharge in a time τ,

$$W = \int_0^\tau I(t)\, V(t)\, dt, \qquad (1.3)$$

and the inductance of the contracting pinch at any instant of time,

$$L(\tau) = \frac{c^2 \int_0^\tau V(t)\, dt}{I(\tau)}, \qquad (1.4)$$

where c is the velocity of light.

Equation (1.4) holds if the real resistance of the pinch can be neglected compared with the inductive reactance. This condition is generally satisfied for powerful pulsed discharges of short time duration. Assuming that there is a strong skin effect and using the formula for the inductance of a coaxial line $L = 2l \ln(r_1/r_2)$, we can construct a curve showing the variation of the effective plasma radius as a function of time.

The method of determining the radius of the plasma column using current and voltage curves is used widely for investigating the dynamics of short pulsed discharges [1, 2]. The data obtained by other methods, in particular magnetic probes and high-speed photography, indicates that the accuracy in the determination of the effective radius of the discharge remains high during all stages of contraction. The good agreement between these results indicates the small contribution of the ohmic resistance to the total resistance of the plasma column. However, in later stages, after maximum contraction, the development of various hydrodynamic instabilities can disturb the axial symmetry of the discharge and the effective radius obtained from the current and voltage curves can be much too low.

Equations (1.3) and (1.4) have been written under the assumption that the plasma resistance R_{pl} is small compared with its inductive reactance $\frac{1}{c^2 I} \cdot \frac{d}{dt} (LI)$. It is also interesting to consider the other limiting case $\frac{1}{c^2} \cdot \frac{d(LI)}{dt} \ll RI$, which occurs in devices in which the plasma column is stabilized by an external magnetic field, for example, in the stellarator. If the plasma configuration remains essentially unchanged in a smooth variation of the current $(dI/dt \approx 0)$ then by measuring the voltage drop and the magnitude of the discharge current it is possible to determine the conductivity of the plasma and, thus, its electron temperature. The electron temperature is related to the conductivity by the following expression [3]:

$$\sigma = 1.5 \cdot 10^{-4} \frac{T_e^{3/2}}{Z \ln \Lambda} \ \Omega^{-1} \cdot cm^{-1}. \tag{1.5}$$

Here T_e is the electron temperature in degrees Kelvin; Z_e is the ion charge; $\ln \Lambda$ is the Coulomb logarithm, which is usually taken to be 10-15:

$$\Lambda = \frac{3}{2e^3} \left(\frac{k^3 T^3}{\pi n_e} \right)^{1/3}, \tag{1.6}$$

where n_e is the electron density.

The electron temperature determined from the conductivity sometimes exceeds the true plasma temperature because of the presence of runaway electrons, which can carry an appreciable part of the plasma current. One of the signs of runaway electrons is intense x-ray radiation, with energies appreciably greater than the mean energy of the plasma electrons. Another common source of error in the determination of electron temperature from the conductivity of a hydrogen plasma derives from the neglect of impurities. Even at relatively low electron temperatures, under the conditions that obtain in the establishment of ionization equilibrium the effective charge of the impurity ions can be rather high; hence, a small admixture in a hydrogen plasma can cause the apparent electron temperature to be high by a factor of 1.5−2. Finally, a third source of error is the neglect of the interaction between the electrons and fluctuations of the electric field that appear as the result of various instabilities (the so-called anomalous resistance of a turbulent plasma [4]).

In addition to methods based on oscilloscope traces of the discharge current and the circuit voltage for determination of the conductivity of a cold plasma, frequent use is made of measurements at frequencies appreciably greater than the characteristic frequency for the development of the discharge. The method of measuring the conductivity at a high frequency in toroidal devices and stellarators [5] is used in the investigation of afterglow plasmas, in which case the discharge current becomes vanishingly small. For this purpose a small toroidal coil supplied by a high-frequency current is mounted on the vacuum chamber in such a way that the plasma pinch threads the coil. Then the closed pinch acts as the secondary winding of a transformer; the primary winding is the toroidal coil. The high-frequency current in the plasma is determined by the expression

$$I_\sim = \frac{V_\sim}{n_1 R_{\mathrm{pl}}}, \tag{1.7}$$

where n_1 is the number of turns of the toroidal coil and $V_\sim$ is the high-frequency voltage applied to the coil. By measuring the current $I_\sim$ in the plasma by means of a Rogowsky loop it is easy to obtain the resistance of the pinch. It should be emphasized the method of measuring conductivity at high frequencies can be used only in very cold plasmas, in which the ohmic resistance exceeds the inductive reactance even at the high frequencies.

The basic difficulty in the determination of electron temperature by the measurement of resistance lies in the need for having
additional information on the geometry of the pinch. The problem
is especially complicated for a plasma whose boundaries are not
sharply defined, in which case the value of the electron temperature is sensitive to the choice of the effective cross section of
the plasma.

The effective radius of a current-carrying pinch is usually
determined by magnetic probes. Under clean vacuum conditions,
with a relatively weak interaction between the discharge and the
walls of the chamber, magnetic probes are not usually used, since
the introduction of probes will have a strong effect on the plasma
behavior. The effective radius of the discharge can be obtained
from the inductance of the discharge, as is done in high-speed
intense pulsed discharges. However, under typical conditions in
toroidal chambers the measurement of the inductance is made
difficult by the fact that the inductive reactance and the real resistance of the plasma are comparable in magnitude; moreover,
each of these components changes drastically during the discharge.

In order to circumvent these difficulties it is necessary to
measure the inductance at a frequency ω which is much greater
than the frequency of the basic discharge, in which case the inductive reactance is much greater than the real resistance [6, 7].
The amplitude of the high-frequency voltage is taken to be as
small as possible so that the perturbation of the primary discharge current will be a minimum. In experiments on the Tokamak
device the high-frequency component of the voltage around the loop
is produced by connecting in the excitation circuit of the discharge
an additional tank coil with an inductance that is much smaller
than that of the primary coil.

The results of such measurement can be interpreted most
simply for an axially symmetric pinch which is not displaced very
far from the axis of the chamber. If the frequency of the additional circuit voltage is such that there is a clearly defined skin
effect, the radius of the pinch is related to the inductance by the
approximate formula

$$L \approx 2l \ln \left(\frac{r_\kappa}{r_{pl} - \dfrac{c}{\sqrt{8\pi\sigma\omega}}} \right). \tag{1.8}$$

References

1. L. A. Artsimovich, Controlled Thermonuclear Reactions, Pergamon Press, 1965.
2. A. M. Andrianov et al., Plasma Physics and the Problem of a Controlled Thermonuclear Reaction, Pergamon, New York, 1958, Vol. 2.
3. L. Spitzer, Jr., Physics of Fully Ionized Gases, Interscience, New York, 1965.
4. A. A. Vedenov, Plasma Physics, Izd. INI AN SSSR, Moscow, 1959.
5. T. Coor et al., Phys. Fluids., 1:411 (1958).
6. N. V. Donskoi, Zh. Tekh. Fiz., 32:1095 (1962) [Sov. Phys. — Tech. Phys., 7(9):805 (1963)].
7. D. P. Ivanov and S. S. Krasil'nikov, in: Plasma Diagnostics, Gosatomizdat, Moscow, 1963, p. 292.

Chapter 2

Plasma Diagnostics with Probes

§ 2.1. Langmuir Probes

Langmuir probes have been used widely in the measurement of plasma parameters in gas discharges for almost forty years [1]. Although Langmuir probes have been described in hundreds of papers, at the present time there is still not available a general theory that makes it possible to interpret probe characteristics under arbitrary conditions. The difficulties are especially marked when probes are used to determine the parameters of a plasma in a magnetic field.

We shall start the present discussion by discussing briefly some of the general considerations that apply to the use of probes in steady-state plasmas with no magnetic field. The presence of even a small conducting body in a plasma invariably leads to perturbations of the plasma because the previously uniform plasma now contains a surface at which ion recombination can occur. In the general case the velocities of the negatively and positively charged plasma particles are different; hence the probe, which is at the plasma potential, will generally have incident upon it more of the fast particles. The higher velocities are almost always associated with the electrons, so that a probe at the plasma potential receives a negative current.

It is obvious that any change in potential will change the current in the probe circuit, this current representing the algebraic sum of the currents due to the flows of positive particles and negative particles. Depending on the sign of the probe potential

with respect to the plasma, the electric field will tend to inhibit
the acceptance of particles of one sign or the other. The region
in which the electric field is concentrated is usually called a double
sheath. The theory of Langmuir probes is based on the assumption that the charged particles within the double sheath move under the influence of the electric field with no collisions. In other
words, the applicability of the theory of probes is limited by the
values of the density and temperature of the plasma since it is
necessary that the mean free path of the particles be greater than
the thickness of the double sheath. When the currents to the probe
are small the thickness of the double sheath is given by the Debye
radius

$$\lambda = \sqrt{\frac{kT}{4\pi n e^2}} = 7.5 \cdot 10^2 \sqrt{\frac{T(\text{eV})}{n\,(\text{cm}^{-3})}}. \qquad (2.1)$$

Here, T is the temperature and n is the density of the plasma. At
large values of the current the thickness of the double sheath increases and can be determined from the well-known Langmuir
relation

$$I = 2.4 \cdot 10^{-6} \frac{V^{3/2}}{x^2} \cdot S, \qquad (2.2)$$

where I is the current to the probe, in amperes; x is the thickness
of the double sheath, in cm; S is the collecting surface of the
probe, in cm^2; V is the potential difference between the probe and
the plasma, in volts. The thickness of the double sheath is determined by the larger of the two values given by Eq. (2.1) and Eq.
(2.2).

Almost the entire potential difference between the plasma and
the probe is concentrated in the double sheath and only a small part
of it, $\sim kT_e/2e$, penetrates into the plasma. The ions that arrive at
the double sheath are accelerated by this potential difference and
this phenomenon is very important in a plasma in which the ion
temperature is low compared with the electron temperature ($T_e \gg$
T_i). In this case the ions arrive at the double sheath near the probe,
which is at a negative potential, with the velocity they would have
if the ion temperature were equal to one-half the electron temperature.

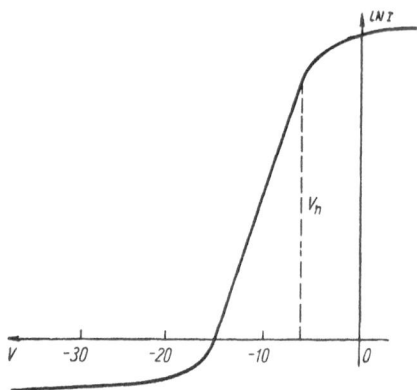

Fig. 2.1. Volt-ampere characteristics of a
probe.

A typical volt-ampere characteristic for a probe, plotted on a
semilogarithmic scale, is shown in Fig. 2.1. At a large negative
potential ($| V | \gg kT_e$) the current to the probe is essentially in-
dependent of the voltage. The negative potential prevents the acceptance of electrons by the probe and the current is determined
by the flux of positive ions to the collecting surface. With an isotropic velocity distribution the positive-ion saturation current is
given by

$$I = \frac{env_i S}{4} \,.$$ \hfill (2.3)

This expression holds only when the electron temperature is
lower than the ion temperature. If this situation does not obtain
the ions arrive at the probe with a velocity that corresponds to
approximately one-half the mean energy of the electrons. According to the calculations given by Bohm [2], when $T_i \ll T_e$ the saturation current is

$$I = 0.4 \, enS \sqrt{\frac{2kT_e}{M}} \,.$$ \hfill (2.4)

Here, S is the collection surface of the probe and M is the ion
mass. If the thickness of the double sheath is much less than the
probe dimensions the area of the collecting surface is equal to the
area of the probe. In a low-density plasma, in which the Debye

radius is equal to or greater than the probe size, the exact defini-
tion of S becomes somewhat complicated. The area of the collecting surface in a low-density plasma can be determined from
the numerical calculations given by Chen [3].

As the probe potential is increased with respect to the plasma
the surface will start to receive more and more of the plasma
electrons: first the fast electrons, and then the slower electrons.
As a result the positive current to the probe is reduced. The current vanishes when the electron and ion flows to the probe are
equal. The potential at which the total current vanishes is usually
called the floating potential. This is the potential acquired by an
isolated body in a plasma. The magnitude of the floating potential
is determined primarily by the electron temperature,

$$V_{\mathrm{pl}} = \frac{kT_e}{2e} \ln \frac{M}{m} \cdot \frac{T_e}{T_i}. \tag{2.5}$$

As the probe potential is increased still further the current becomes negative and the magnitude of the current tends to increase
with the potential. At potentials that correspond to this region of
the characteristic the collecting surface can collect electrons
whose energies exceed the value V_e, where V_e is the retarding
potential between the plasma and the probe. In addition to the flow
of electrons to the probe, there is still a flow of positively charged
ions; hence, in determining the magnitude of the electron current
it is necessary to correct for the ion saturation current. If the
electrons exhibit a Maxwellian energy distribution the current
density to the probe at a retarding potential V will be $e^{-eV/kT}$
times smaller than the electron saturation current density, that
is to say, the current density to the probe in the absence of a retarding field. Thus, with a negative current to the probe the electron current is determined from the following relation:

$$I = \frac{1}{4} e n \bar{v}_e S e^{-\frac{eV}{kT}}, \tag{2.6}$$

where

$$\bar{v}_e = \sqrt{\frac{8kT_e}{\pi m}}.$$

With a further increase in the potential of the probe with respect
to the plasma there appears a rather sharp break in the character-

istic and the current to the probe becomes relatively independent
of the potential. At the break point the assumptions used to de-
rive Eq. (2.6) are no longer valid, that is to say, the retarding
potential vanishes. The increase in electron saturation current
with further increases in potential is then a consequence of the
increase in effective collecting surface of the probe caused by the
increasing thickness of the double sheath.

We turn now to a consideration of the determination of plasma
parameters from an analysis of probe characteristics. Probe
measurements can be used to determine the plasma potential at a
given point, the ion density (which to a high degree of accuracy is
equal to the electron density), and the electron temperature. The
electron temperature can be obtained most conveniently from a
plot of the probe characteristics on a semilogarithmic scale (cf.
Fig. 2.1); on this curve one observes an extended straight seg-
ment. Analyzing Eq. (2.6) we find

$$\ln I_e = \ln \left(\frac{1}{4} \bar{v}_e \, enS \right) - \frac{eV}{kT_e}. \tag{2.7}$$

Consequently, the tangent to the probe characteristic on a
semilogarithmic plot is e/kT_e. In order to determine electron
density (which is equal to the ion density by virtue of the plasma
neutrality condition) use can be made of Eq. (2.3) or Eq. (2.4), depending on the relation between the electron temperature and the
ion temperature. The ion saturation current is measured with a
large negative bias, in which case the volt-ampere characteristic
reaches a plateau. Physically, the saturation effect means that the
negative bias of the probe potential with respect to the plasma is so
large as to completely prevent electrons from reaching the probe.
In order to determine the electron density it is necessary to know
the area of the collecting surface, which increases with increasing
retarding potential; hence, even at large potentials the current to
the probe will increase somewhat with retarding potential. The
collecting surface is located at a distance x from the surface of
the probe and can be computed from Eqs. (2.1) or (2.2). In those
cases in which the density must be determined with the highest
possible accuracy it is necessary to eliminate the effect of the
thickness of the double sheath and the surface of the conducting

insulator from the results of the measurements. For this reason in reference [4] a flat collecting surface with a guard ring was used. The guard ring is kept at the same potential as the collecting surface. In a probe of this design, the increase in the thickness of the double sheath moves the collecting surface away from the surface of the probe, leaving the shape and dimensions of the former unchanged. The saturation of the ion current in the presence of the guard ring occurs at a much lower potential than in the absence of the ring and a further increase in potential has essentially no effect on the current to the probe.

In addition to being used for the determination of electron temperature and density, Langmuir probes are used for measuring the potential of a plasma with respect to electrodes or chamber walls. As we have indicated above, when the potential of the probe with respect to the plasma is zero the electron branch of the probe characteristic (plotted on a semilogarithmic scale) exhibits a break due to the fact that the exponential dependence of the probe current holds only for a negative probe potential with respect to the plasma (cf. Fig. 2.1). Unfortunately, the break in the probe characteristics is rather weak, especially in the presence of a magnetic field, so that the accuracy in the determination of plasma potential by this method is rather poor.

More exact measurements of plasma potential can be obtained by the use of a so-called hot probe, which was also first proposed by Langmuir. The principle of the hot probe is very simple. The probe itself is a thin tungsten wire. The probe characteristics are taken twice. One measurement is taken with current flowing through the wire; this current heats the wire to a temperature at which a significant thermionic emission occurs. In the second measurement the probe characteristic is taken with the wire cold. At positive probe potentials the temperature of the wire will not have any effect on the characteristic since the cold emission electrons are retarded by the field between the probe and the plasma. However, even with small relative potentials of the probe with respect to the plasma the thermionic electrons escape from the probe into the plasma; as a result, the characteristics taken with the wire hot and with the wire cold no longer coincide. The probe potential at which the curves diverge is then equal to the plasma potential.

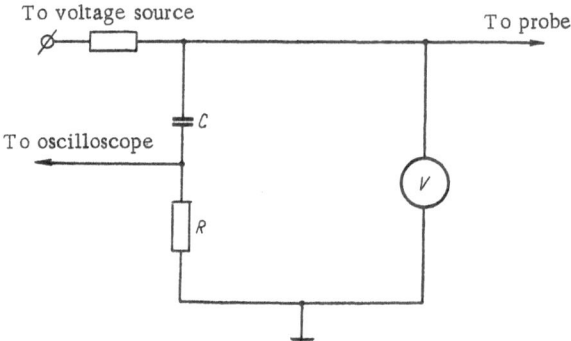

Fig. 2.2. Circuit for measuring probe current with a
fixed bias.

Errors in the determination of the plasma potential at small
values are due primarily to the voltage drop on the tungsten wire
itself. At large plasma potentials the error due to this effect is
unimportant. Absolute values of the potential measured in this
way can reach several thousands of volts [5].

In thermonuclear research the primary concern is with plasmas whose lifetimes are much shorter than a second. In taking
the probe characteristics in such plasmas an oscilloscope can be
used in the following way: a signal proportional to the probe current is applied to one pair of plates; the other pair of plates receives a voltage from a saw-tooth generator. This voltage is applied between the chamber walls and the probe. It will be obvious
that oscilloscope traces of probe characteristics are convenient
for determination of density and temperature under conditions in
which the length of the saw-tooth pulse is smaller than the characteristic time for a change in the plasma parameters. The oscilloscope method makes it possible to obtain a probe characteristic in a pulsed plasma device in a single discharge.

Probe characteristics in a plasma in a pulsed device that has
good reproducability can be obtained in a simpler manner. For
this purpose we take an oscilloscope trace of the current to the
probe for a fixed probe potential with respect to the chamber walls.
By taking oscilloscope traces at different probe potentials it is
possible to obtain the data needed for plotting the probe characteristics corresponding to any point in the process. A simple

scheme for recording probe current is shown in Fig. 2.2. The time
constant of the RC circuit must be much greater than the plasma
lifetime if the probe potential is to remain unchanged in the measurement process. It is also necessary that the voltage drop across
the resistance R have no effect on the probe potential. In order to
obtain the data properly by this method it is necessary to maintain
a constant plasma potential and to avoid electric arcs between the
probe and the plasma. The existence of an arc is easily detected
by an anomalously high probe current and by the fact that the surface of the probe becomes hot enough to be visible. The production of arcs sets a limitation on the applicability of the probe
method. This limit appears at densities of 10^{14}-10^{15} cm^{-3}.

The use of probes at high temperatures is limited by secondary emission effects. The limit lies in the region of electron
temperatures of several tens of electron volts. Frequently, especially in the presence of an external magnetic field (in which
the interpretation of the probe characteristics is complicated),
probes are also used for measuring the density using the saturation ion current. For this purpose a large known negative potential is applied to the probe in order to eliminate the effect of the
plasma potential on the measured results. Because of the weak
dependence of the saturation current on the plasma temperature,
an oscilloscope trace of the saturation current gives the density
as a function of time with an accuracy that is usually satisfactory.

The presence of a strong magnetic field in a plasma has a
very important effect on the operation of the probe and interpretation of the results requires an analysis of the experimental conditions in each particular case. However, if the magnetic field
is weak and if the radius of curvature of the electron trajectories is significantly larger than the characteristic dimensions
of the probe, the method of analysis of the probe characteristics
is essentially the same as in the absence of a magnetic field. In
a strong magnetic field the probe measures the "longitudinal"
plasma temperature* and its collecting surface is given by the
area of the transverse cross section, perpendicular to the lines of
force of the field. Since the angular distribution of the vector ve-

---

*The longitudinal temperature is taken to mean the temperature which corresponds
to the energy distribution of the velocity component directed parallel to the lines
of force of the magnetic field.

locities in a magnetic field is almost never isotropic, the longitudinal temperature as measured by the probe can be appreciably different from the "transverse" temperature.

In cases in which the plasma potential changes appreciably during the time required for obtaining the probe characteristics it is found that a single Langmuir probe cannot be used: in these cases the so-called double floating probe [6, 26] can be used.

Let us consider two identical isolated probes of area S which are connected through a small resistance R and located in a plasma. At the initial time, before the establishment of the floating potential, each probe is subject primarily to electron current; in the steady-state regime, however, the current to each probe vanishes. Physically this means that the electron current is equal to the ion current. It is then obvious that if there are no gradients of potential and temperature (due to the probe) the plasma current through the resistance R will vanish. The potential distribution for this situation is shown in Fig. 2.3a. The ion current (and the electron current, which is equal but opposite in sign) is given, in accordance with Eq. (2.4), by $i_p \sim ns\sqrt{T_e}$.

Assume now that a small potential difference V is applied between the probes, but that the entire electrical probe circuit is isolated as before. Then, because of the high electron velocity, the potential of both probes will be established below the plasma potential. If the potential of at least one of the probes were higher than the plasma potential electron current would flow to this probe and the magnitude of this current would be equal to the saturation current. The electron saturation current in an isothermal plasma ($T_i \sim T_e$) exceeds the ion saturation current by a factor $(M/m)^{1/2}$ and this means that the electron saturation current cannot be balanced by the ion flow at both probes. Consequently, in the stationary state, in which the algebraic sum of the currents to both probes is equal to zero, the distribution of the potentials must be such that the probe having the higher potential will be at some negative potential with respect to the plasma (cf. Fig. 2.3b). The difference of potentials between the probe and the plasma is limited by the electron current to this probe. The surplus of electron current over ion current to one of the probes is established by the surplus of ion current over electron current to the second probe since the potential barrier that retards the electrons must be

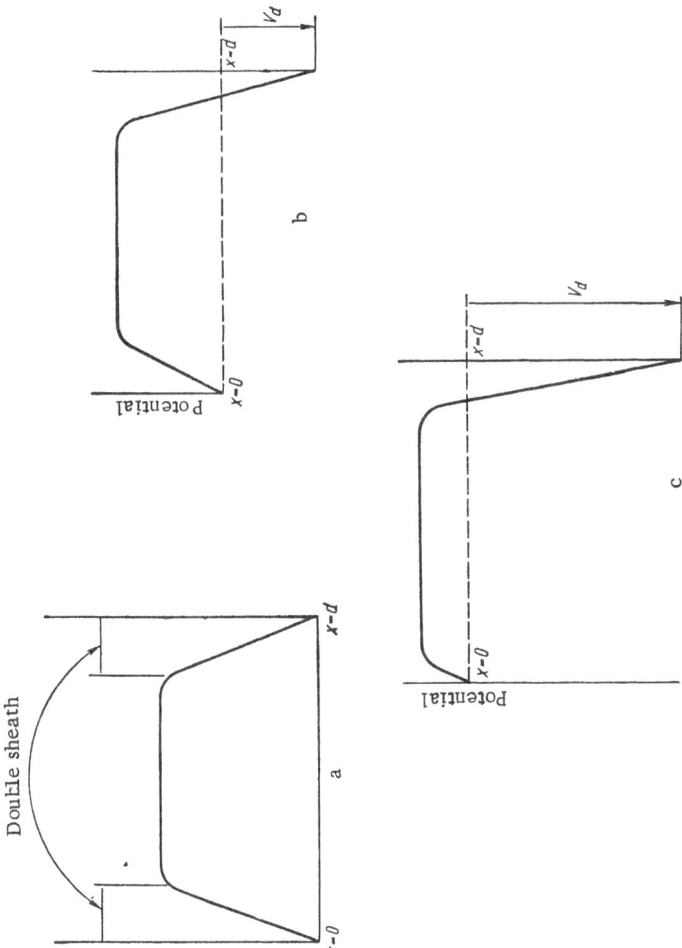

Fig. 2.3. Potential distribution near a double probe. a) Zero potential difference between probes; b) small potential difference between probes; c) large potential difference between probes.

higher here. As a result, through the resistance R there will
flow a current I. In this regime, as a consequence of the negative
bias of both probes with respect to the plasma, the magnitude of
the ion current to both probes is equal to the ion saturation current i_p.

Evidently, the maximum current I for a sufficiently large
potential difference between the probes is given by i_p. Actually,
the electron current is essentially zero to the probe, which is at a
large negative potential with respect to the plasma under conditions of a sufficiently high potential difference between the probes;
thus, the potential of the second probe is such that the electron
current is equal to twice the ion saturation current, i.e., $2i_p$.
Consequently, a current $I = i_p$ flows in the probe circuit.

We now consider quantitatively the current to the double probe
for various potential differences between the probes. The following notation is used: i_1^- and i_2^- are the electron currents to the
first and second probes respectively; i_1^+ and i_2^+ are the ion currents.

The condition that the total current to the probes be zero is

$$i_1^- + i_2^- = i_1^+ + i_2^+. \tag{2.8}$$

Using the fact that the ion saturation current is independent of the
potential difference V between the probes, and assuming a Boltzmann distribution, we obtain the following expressions for the ion
and electron currents:

$$i_1^+ = i_2^+ = i_p = \frac{nev_i S}{4}, \quad i_1^- = i_{10}^- = e^{-\frac{eV_1}{kT_e}},$$

$$i_2^- = i_{20}^- e^{-\frac{eV_2}{kT_e}}, \tag{2.9}$$

where $V_1 - V_2 = V$; the notation is given in Fig. 2.4. Substituting
Eq. (2.9) in Eq. (2.8) and introducing the notation $\Gamma = (2i_p/i_2) - 1$,
we obtain the following result, which is suitable for graphic interpretation of the data:

$$\ln \Gamma = -\frac{eV}{kT_e}. \tag{2.10}$$

At high values of V the exponential nature of the functional dependence no longer holds and i_2^- approaches a limit given by $2i_p$.

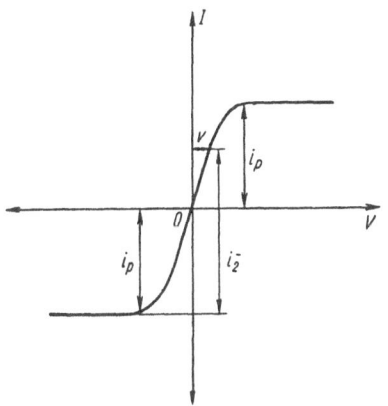

Fig. 2.4. Volt-ampere characteristic of
a double probe.

Physically this means that the electron current to one of the probes is equal to zero while the electron current to the second probe is equal to the sum of the ion currents to both probes. When identical probes are used, owing to the symmetry a change in the sign of V leads to a probe curve which can be obtained from the original curve by mirror reflection with respect to the origin; the saturation current for both branches of the characteristic is given by i_p. A typical probe characteristic is shown in Fig. 2.4. Along the ordinate axis we have plotted the current that flows between the probes I. To avoid confusion we note that the electron current to the probe which is at the smaller negative potential i_2^- is made up of the current in the probe circuit and the ion saturation current, i.e., $| i_2^- | = | I | + | i_p |$.

If the plasma exhibits a potential drop V_x, then the characteristic is no longer symmetric with respect to the origin and Eq. (2.10) assumes the form

$$\ln \Gamma = -\frac{eV}{kT_e} + \frac{eV_x}{kT_e}. \tag{2.11}$$

In making systematic measurements, in which the accuracy of the probe operation is continually verified and no doubt is raised, it is not necessary to compare the theoretical and experimental characteristics for each measurement. It is sufficient to determine T_e by measuring the saturation current and the slope of the characteristic at the origin:

$$T_e = \frac{i_p}{2\frac{dI}{dV}\Big|_{I=0}}.$$

Here T_e is the electron temperature in eV, I is the current in amperes, and V is the voltage in volts.

Since the use of a double probe is based on the absence of a metal connection between the measurement circuit and ground, or elements which have a metal connection to the ground, the use of this probe introduces some additional difficulties. An ingenious solution of the problem is given in [7], where the current flowing between the probes is measured with an oscilloscope through an isolation transformer with a ferrite core.

The analysis of probe characteristics given in the present section is based on the assumption that the particles exhibit a Boltzmann distribution in the double sheath. This assumption is valid when the mean free paths of the particles are much longer than the double sheath, that is to say, when the conditions are such that the diffusion approximation need not be used. The mean free path of a charged particle,

$$\lambda = 1.5 \cdot 10^{12} \frac{T^2 (\text{eV})}{n},\qquad\qquad (2.12)$$

is significantly greater than the dimensions of the double sheath for a wide range of plasma densities and temperatures. When the plasma density is increased the results of probe measurements are usually distorted by the fact that breakdown can occur between the probe and the plasma before the mean free path of the charged particles becomes comparable with the thickness of the double sheath.

§ 2.2. Electrostatic Plasma

Analyzer

Analysis of the characteristics of single and double probes can be used to obtain, with reasonable accuracy, the electron temperature and the plasma density if the temperature is below 20-30 eV and if the density is below 10^{14}-10^{15} cm^{-3}. We find, however, that the ion temperature cannot be obtained by analysis of the probe characteristics. In recent years, a number of laboratories have developed a multielectrode probe for the measurement of ion temperature; this device is frequently called an electrostatic plasma analyzer [8-13]. Energy analysis of the particles in an electrostatic analyzer is realized by means of a retarding field, a technique which has been well known since the time of the Franck–Hertz experiments.

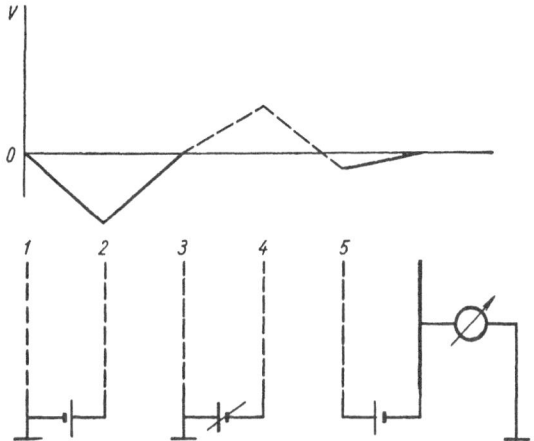

Fig. 2.5. Potential distribution at the grids of an electrostatic analyzer.

Let us consider one of the most popular designs for an electrostatic analyzer intended for the measurement of ion energy distributions (Fig. 2.5). Along the path of the plasma, which passes
through an aperture in a small metal box, there is located a series
of metal grids which are maintained at various potentials. Sometimes, in place of the aperture, use is made of a channel with a
large ratio of length to diameter; this arrangement serves to col-
limate particles with one nonvanishing velocity component. Grids
1 and 2 are designed to separate the electron and ion components
of the plasma and to cut off the electrons. The electron component
can only be cut off if the electric field penetrates into the plasma.
For this reason the cell size of the first two grids must be smaller
than the thickness of the sheath region between the electrode and
the plasma.

In the analyzer designs that are presently in use successful
use has been made of grids with cell size of the order of 20-30 μ.
By using these grids it is possible to operate in plasmas with densities up to 10^{13} cm^{-3} at ion temperatures up to approximately 100
eV. Grids 3 and 1 are usually maintained at zero potential. In
certain analyzer designs these grids are not used at all. The
purpose of grid 3 is to reduce the coupling of the potentials between grids 2 and 4. A positive potential V is applied to grid 4 in
order to retard ions whose directed energy does not exceed the

value 1 eV. Finally, grid 5, which is located directly in front of the collector, serves to eliminate effects due to secondary emission. This grid is maintained at a small negative potential. If this grid is not used there can be considerable error in the measurement of the ion velocity spectrum.

The electrostatic analyzer can be used for measurements of ion spectra close to the walls of the chamber and for measurements of ion spectra far from the walls. In the second case the metal box is introduced into the plasma although the introduction of such a metal surface undoubtedly has an effect on the plasma parameters. When the retarding field method is used for the determination of the energy spectra of weak ion streams the gain of the amplifier is not always sufficient for measurements of the collector current for all values of the retarding potential. Small currents can sometimes be measured with an electron multiplier instead of by direct measurement of the collected current. The particles from the analyzer move through the last grid, beyond which there is a region of accelerating electric field. The ions that leave the analyzer are accelerated by a potential difference of several kilovolts and are then detected by an electron multiplier with a covered cathode.

The use of the retarding field method for investigating a plasma in a magnetic field (this is the most popular case) usually introduces additional difficulties. If the energy of the particles being analyzed is such that their radius of curvature in the magnetic field is of the order of, or smaller than, the distance between grids, a probe with an analyzing electric field can only be used if the grid planes are perpendicular to the lines of force of a magnetic field. In this case the measurement determines the velocity distribution parallel to the lines of force. For any other analyzer orientation the particle motion occurs in crossed fields and the analysis of the volt-ampere characteristics becomes essentially impossible. Measurements of the energy of circular particle motion are possible only when the radius of curvature of the particle trajectory is significantly greater than the characteristic dimensions of the analyzer.

We now consider a typical analysis of results obtained with a retarding-field electrostatic analyzer. The negative bias on the second grid (with zero potential on the fourth grid, which serves to

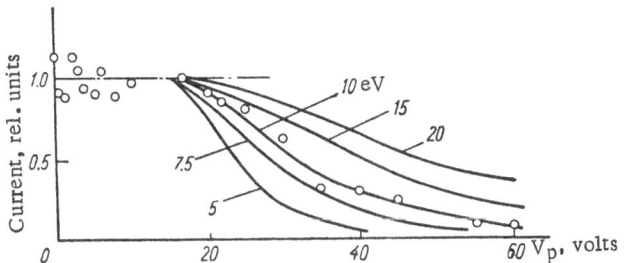

Fig. 2.6. Volt-ampere characteristics corresponding to
various temperatures and the experimental points.

produce a retarding field for the ions) is chosen in such a way as
to obtain the saturation current of the ions that strike the collector. The ion saturation current corresponds to total cutoff of the
electrons. In this situation the collector current is measured as
a function of the positive potential on the fourth grid. A typical
functional relation that is obtained as a result of such measurements
is shown in Fig. 2.6 [9]. It is evident from the figure that the ion
current is essentially unchanged with a retarding voltage up to 20
V, that is to say, the energy of all ions passing through the grid
exceeds 20 eV. Physically this means that the plasma potential
differs from the zero potential which is maintained on the first
grid (in the present experiments the first grid was maintained
at the potential of the chamber walls) and is equal to 20 V. Then,
as the retarding potential is increased the magnitude of the recorded ion current is reduced.

   If the velocity distribution of the particles is known it is not
difficult to write down an expression for the experimentally obtained relationship. If one deals with a collimated beam directed
perpendicularly to the plane of the grids and if a retarding potential
+V is applied to the fourth grid the ion current to the collector
(under the assumption of a Maxwellian velocity distribution) is

$$j = \int_{v_1}^{\infty} ev\,dn = 2en \sqrt{\frac{2kT_i}{\pi M}} \left(1 + \frac{eV}{kT_i}\right) e^{-\frac{eV}{kT_i}}, \qquad (2.13)$$

where

$$v_1 = \sqrt{\frac{2eV}{M}} .$$

Calculated curves are shown in Fig. 2.6 for temperatures of
5, 7.5, 10, 15, and 20 eV; these curves indicate the accuracy of
the method. All the curves are normalized to the experimental
value of the ion current at the point V = 0.

§ 2.3. Magnetic Probes

and Pressure Balance

In all thermonuclear devices based on the notion of passing
large currents through a plasma, and in magnetic traps in which the
dimensionless parameter $\beta = nkT/H^2/8\pi$ is reasonably large, the
basic measurement instrument is the magnetic probe. Indeed, it is difficult to picture the present state of the art in research on self-
sustaining columns with large discharge currents or the investigation of the interaction between plasmoids and the magnetic field
without magnetic probes.

The principle of the magnetic probe is extremely simple. A
coil is oriented in a definite way at the point at which one wishes
to obtain the time variation of the magnetic field. Very frequently
this point is in a region filled with plasma; hence, in making measurements with electrostatic probes it would be necessary to have
data on the effect of the measuring instrument on the object being
investigated. It is possible to verify the effect of the instrument,
as a rule, only by some sort of substitution experiment. A simple
control experiment can be carried out by introducing two identical probes into the plasma and noting the dependence of the probe
signal as a function of distance between the probes. It will be
evident that the smaller probes cause smaller perturbations. At
the present time a systematic investigation of the perturbations
that are introduced by probes is not available. In [14] some attention is given to the distortion of the current distribution in the
plasma caused by the introduction of a probe. For the planar case
with a strong skin effect the distortion of the magnetic field at
distances equal to the probe diameter is found to be 10%.

The perturbation of the current through a plasma caused by a
probe has been considered in greater detail in [15], in which it
is shown that errors in the probe measurements become significant
in the determination of the magnetic field distribution close to the

axis of the discharge. This result is to be expected because in the case of axial symmetry the magnetic field due to the current is zero at the axis and small perturbations introduced by the probe can lead to erroneous readings of large values of $H_\varphi(r)$ near to the point $r = 0$; in turn these can lead to an erroneous conclusion indicating a sharp current maximum at the axis. However, the principal effect of a probe on a plasma is usually not distortion of the current distribution, but rather contamination due to evaporation of the probe wire. Limitations caused by the evaporation of the wire are discussed at the end of the present section.

The most popular magnetic probe is a small coil whose linear dimension is approximately one millimeter and that contains several tens of turns on a small ceramic tube. In order to suppress electrostatic noise a thin-walled metal tube is usually located inside the ceramic tube. The wall thickness of the shielding metal tube must be much smaller than the skin depth at the highest rate of change of the field being recorded by the probe. The signal from the probe goes to an integrating circuit and, after amplification, to the oscilloscope. It is obvious that the oscilloscope records the voltage from the probe, which is proportional to the derivative of the magnetic flux. To obtain a curve showing the field itself it is necessary to carry out a graphical integration.

By making a series of measurements with the probe located at different points in space it is not difficult to obtain curves showing the spatial distribution of magnetic field at different instants of time. However it is not always possible in practice to plot curves of the current density in the plasma on the basis of the data obtained in this way.

In this respect, it is fortunate that in thermonuclear research one frequently encounters axially symmetric field distributions. For example, in intense short-lifetime pulsed discharges which are not stabilized by external fields, the axial symmetry is maintained over the entire inertial compression stage. In this case the current density vector is parallel to the axis of the discharge and the axial component of the magnetic field H_z vanishes, so that the equation

$$\text{curl } \mathbf{H} = \frac{4\pi}{c} \mathbf{j} \qquad (2.14)$$

is given by

$$j_z = \frac{c}{4\pi r} \cdot \frac{\partial}{\partial r} (rH_\varphi) \qquad (2.15)$$

in cylindrical coordinates.

As a result, the problem of plotting the spatial distribution of the current density reduces to the experimental determination of the curve $H_\varphi(r)$ by means of magnetic probes.

The distribution of current density has been investigated in detail in a powerful pulsed discharge with currents of the order of hundreds of kiloamperes [16-18]. At the beginning of the discharge, during the inertial compression stage of the plasma column, the current distribution exhibits axial symmetry; hence in plotting a histogram of the current density it is sufficient to record the radial distribution of the azimuthal component of the magnetic field.

As we have noted above, the section of the curve corresponding to the region directly at the axis of the discharge is the most difficult to measure. The experimental curve will be a good approximation to the true curve if the value of the current density at the axis is determined by extrapolation of the derivative $\partial H_\varphi / \partial r$ to the point r = 0, making use of the approximation $\partial H_\varphi / \partial r \rightarrow H_\varphi / r$. In this way, using Eq. (2.14) we can obtain the current density at the axis

$$j_z = \frac{c}{2\pi} \cdot \frac{\partial H_\varphi}{\partial r} . \qquad (2.16)$$

The distribution of discharge current density can also be investigated by means of small Rogowsky loops which are introduced into the discharge. However, since these are large as compared with probes, loops will tend to introduce a larger distortion.

In discharges whose lifetimes are significantly greater than the inertial time, in addition to obtaining the current distribution it is also possible to determine the spatial distribution of the gas-kinetic pressure nkT. In this case the pressure balance relation contains two terms: the gas-kinetic pressure and the magnetic pressure. We can write the equilibrium condition for the plasma and the magnetic field in the following form:

$$-\frac{\partial P}{\partial r} = \frac{1}{c} |jH|. \tag{2.17}$$

In the presence of a longitudinal field and axial symmetry, assuming that all quantities are independent of z, we have

$$-\frac{\partial P}{\partial r} = \frac{1}{4\pi} \left[\frac{1}{r} H_\varphi \frac{\partial (rH_\varphi)}{\partial r} + H_z \frac{\partial H_z}{\partial r} \right], \tag{2.18}$$

where H_z and H_φ depend only on r. By using magnetic probes to measure the functional relations $H_z(r)$ and $H_\varphi(r)$ at different instants of time and by carrying out the required mathematical operations in graphical form, it is easy to obtain the functional relation $P = f(r)$. Similar methods can be used to determine the distribution of gas-kinetic pressure in discharges in toroidal chambers [19].

The determination of the gas-kinetic pressure is not the only problem in plasma diagnositics that can be solved by means of magnetic probes. By analyzing the curves for the distribution of magnetic field obtained by means of magnetic probes it is possible to obtain important information concerning the inductance of the gas discharge, the penetration of the induction electric field, and a number of other important parameters. As an example, it is of interest to consider the possibility of determining the plasma conductivity from probe measurements. We start with a generalization of Ohm's law,

$$j = \sigma \left(E + \frac{1}{c} [vH] \right). \tag{2.19}$$

If it is possible to measure the component of current density and electric field directed along the lines of force, the conductivity of the plasma (more precisely, the value of the conductivity in the absence of a magnetic field $\sigma = 1.9 \cdot 10^2 \frac{T_e^{3/2}(eV)}{z \ln \Lambda}$ $\Omega^{-1} \cdot$ cm^{-1}) is determined from the simple relation

$$\sigma_{\parallel} = \frac{j_{\parallel}}{E_{\parallel}}. \tag{2.20}$$

Consequently, in addition to finding the magnetic field components by using different orientations of the magnetic probes it is also possible to find components of the electric field and the current density.

Under conditions of axial symmetry, for example, in a discharge stabilized by a longitudinal magnetic field H_z, the current density can be determined from the Maxwell equation (2.14) in the following way:

$$j_z = \frac{c}{4\pi r} \cdot \frac{\partial}{\partial r}(rH_\varphi);$$

$$j_\varphi = -\frac{c}{4\pi} \cdot \frac{\partial H_z}{\partial r}; \quad j_r = 0. \tag{2.21}$$

It is evident from Eq. (2.21) that the current density is determined completely by the spatial distribution of the axial and azimuthal components of the magnetic field.

Writing the other Maxwell equation in cylindrical coordinates,

$$\mathrm{curl}\ \mathbf{E} = \frac{1}{c} \cdot \frac{\partial \mathbf{H}}{\partial t}, \tag{2.22}$$

we obtain expressions that make it possible to compute the magnitude and direction of the electric field, using the same curves that give the distribution of magnetic field,

$$E_z(r) = \frac{1}{c} \cdot \frac{\partial}{\partial t} \int_0^r H_\varphi\, dr + E_z(0); \tag{2.23}$$

$$E_\varphi(r) = -\frac{1}{cr} \cdot \frac{\partial}{\partial t} \int_0^r rH_z\, dr. \tag{2.24}$$

A typical example of the accurate analysis of magnetic probe measurements for the purpose of obtaining maximum information on the behavior of a discharge stabilized by an external magnetic field is given in [19-20].

It will be obvious that the possibility of determining the conductivity of a plasma for computing the electron temperature can be realized only in the absence of instabilities. In the presence of instabilities the axial symmetry of the discharge can be distorted and the functional dependence on the plasma parameters can become very complicated.

Magnetic probes have received wide usage in experiments on plasma confinement in magnetic traps. A probe located inside the trap records the reduction of the magnetic flux due to the diamagnetism of the plasma. It is especially convenient to analyze the results of measurements in those regions of the trap in which the lines of force of the magnetic field are straight. The boundary of the plasma is parallel to the lines of force and the magnetic pressure on the plasma is proportional to the square of the strength of the external field. The equilibrium condition at the boundary is

$$nk\,(T_e + T_i) = \frac{H_e^2}{8\pi} - \frac{H_i^2}{8\pi} \, . \tag{2.25}$$

Here H_e is the strength of the external magnetic field while H_i is the strength of the field inside the plasma.

Thus, if the plasma density or temperature is known, the other quantity can be determined from the diamagnetic effect.

The need for the proper positioning of the magnetic probe, so that the probe itself does not cause a strong distortion of the magnetic field, must be considered in greater detail. In considering the various causes of error due to the improper location of the magnetic probe in an axially symmetric adiabatic trap we may consider the following two examples.

1. A uniform plasma column of radius R is located in a constant uniform magnetic field. Inside the plasma the field is given by H_i; outside it is given by H_e. Aligned coaxially with the plasma column there is a ceramic tube of radius r_0 within which there is a measuring coil. Under these conditions the plasma is bounded by two cylindrical surfaces of radii R and r_0. The pressure balance condition (2.25) is satisfied at both surfaces. The equilibrium is established after part of the plasma moves to the wall of the glass tube and the strength of the magnetic field becomes equal to H_e. Consequently, the magnetic probe located inside the tube does not record the diamagnetic signal. If the probe is to record the diamagnetic signal properly, that is to say, the signal corresponding to the field expelled by the plasma, it is necessary that the probe entrance be oriented.

2. A cylindrical vacuum chamber with a glass tube located along the axis is surrounded by a massive conducting chamber.

In the absence of the plasma the strength of the longitudinal magnetic field is H_e, as in the first example. When the chamber is
filled by plasma part of the flux moves out of the region occupied
by the plasma. As a consequence of the skin effect the massive
conducting chamber does not allow immediately the establishment
of the initial value of the field H_e in the gap between the plasma
and the wall so that the strength of the field in the gap is increased.
Since the pressure balance condition (2.25) is satisfied at both
cylindrical boundaries of the plasma $r = r_0$ and $r = R$, the magnetic
probe not only records the reduction of the field due to the diamagnetism of the plasma, but also does not record the increase of
the magnetic field inside the dielectric tube located coaxially with
the plasma column.

If the plasma occupies almost the entire cross section of the
vacuum chamber the pressure of the plasma as averaged over the
cross section can be determined by measuring the emf in a coil
wound around the chamber. It is especially useful to employ
measuring coils that are wound around the region filled with plasma in the experiments on the determination of the velocity of a
plasmoid along the magnetic field; with this method it is possible
to determine the exact time at which the plasmoid passes a given
point as well as its shape, without any direct contact between the
plasma and the measurement device. Since the inductance of a
coil wound around the chamber can be made rather large, the
signal can be derived with a small resistance; in this case, the
output is actually the integrated signal from the probe, that is to
say, the flux change.

It should be noted that when the gas-kinetic plasma pressure
nkT is determined by measurements of the diamagnetic signal there
is a risk of errors associated with the improper use of Eq. (2.25).
If the collision frequency in the plasma is comparable with, or
larger than, the Larmor frequency, the gas-kinetic plasma pressure is in equilibrium not only with the difference in magnetic
pressures inside and outside the plasma, but also the pressure at
the wall; as a result the measured value of H_i obtained by substitution in Eq. (2.25) gives a value of the gas kinetic pressure
that is too low.

The role of magnetic probes in plasma physics is not limited
to measuring magnetic fields in systems designed for plasma con-

finement. Magnetic probes also find wide usage in experiments
on the study of shock waves in magnetoactive plasmas. Such
questions as the determination of the oscillatory structure of the
shock front [21] and the investigation of the nature of the dissipa-
tion of the wave can be studied by means of magnetic probes.
This method of investigation has been a basic one in experiments
on oblique waves [22].

In considering the experimental data obtained with magnetic
probes it is necessary to note their role in the investigation of
interactions between high-frequency fields and a plasma. One of
the most interesting results is the spatial amplification of high-
frequency fields at magnetoacoustic resonances [23]. To obtain
the complete pattern of the distribution of the high-frequency field
the azimuthal and axial components of the high-frequency field
were both determined. The shape of the distribution and the dependence on basic parameters were found to be in good agreement with the theoretical predictions.

An interesting feature of the distribution of the high-frequency
field at a magnetoacoustic resonance is the peak in the intensity
of the field at the chamber axis. This finding means that the fixed
magnetic field which confines the plasma also experiences the
pressure of the high-frequency field (the interaction between the
high-frequency field and the fixed field takes place throughout the
plasma). In this case the pressure balance relation is

$$nk\,(T_i + T_e) + \frac{\tilde{H}^2}{8\pi} + \frac{H^2}{8\pi} = \text{const}, \qquad (2.26)$$

where $\tilde{H}$ is the intensity of the high-frequency field.

Thus, the measurement of the diamagnetic effect at the magnetoacoustic resonance does not give a direct determination of
the gas kinetic pressure. Instead of obtaining the gas kinetic pressure we actually obtain the sum $nkT + \tilde{H}^2/8\pi$. This example
shows once again that the determination of plasma parameters
from the diamagnetic signal always requires careful analysis of
the experimental conditions.

A principal feature of the application of magnetic probes in
the measurement of the distribution of magnetic fields in a plasma
is the unavoidable direct contact between the wall of the probe and

the plasma. Two basic questions arise in this connection:

1. Does the probe immersed in the plasma record all of the changes in the magnetic field?

2. For what plasma parameters does heating of the probe cause evaporation of the material of which the shell is made, that is to say, contamination of the plasma and the destruction of the probe itself?

We shall consider each of these questions separately.

1. If the strength of the magnetic field in the region occupied by the probe is to be tracked successfully during the time in which the magnetic field in the plasma is changing, the characteristic time for the field change must be greater than the time for the magnetic field to diffuse from the plasma into the region occupied by the probe.

The diffusion time for the magnetic field in a cylindrical cavity of radius a can be estimated from the formula for field damping in such a cavity [24],

$$t \approx \frac{\pi \sigma a^2}{c^2}. \tag{2.27}$$

Here σ is the plasma conductivity. Substituting the value $\sigma = 10^{16}$ cgse, which corresponds to an electron temperature of approximately 100 eV, we obtain the resolution time for a probe 2 mm in diameter, which is found to be tenths of a microsecond.

2. The energy carried by the particles that strike the probe is determined by the temperature of the plasma and by the potential difference which is established between the plasma and the insulated probe immersed in the plasma. In turn, this potential difference is determined by the temperature (cf. §2.1). A simple calculation carried out under the assumption of a Maxwellian equilibrium with $T_e = T_i$ can be used to estimate the heat flux per unit surface of the probe [24],

$$Q = \frac{n}{2\sqrt{2\pi M}} (kT)^{3/2} \left(\ln \frac{M}{m} + 8 \right) \text{ erg/cm}^3 \cdot \text{sec}. \tag{2.28}$$

The heating of the probe due to the particle flux is always greater than the heating due to radiation from the plasma. In estimating the thermal effect of the plasma on the probe the thermal isolation due to the magnetic field is not taken into account. The role of the magnetic field can be neglected since there is always a section of the surface of the probe which intersects the lines of force of the magnetic field; under these conditions the magnetic field does not shield the probe from particles. Using the available data for the boiling temperature and the heat capacity of quartz and alumina we can easily obtain relations that enable us to estimate the time required for heating the probe to a temperature corresponding to evaporation of the wall material. The formulas used to estimate the limiting time for a shell made of quartz and alumina are

$$t_{\text{alum}} \approx \frac{4 \cdot 10^{30}}{n^2 T^3(\text{eV})}; \quad t_{\text{quartz}} \approx \frac{4 \cdot 10^{29}}{n^2 T^3(\text{eV})}. \tag{2.29}$$

In writing Eq. (2.29) we have not taken account of heat removal through the feedthroughs, etc.

Contactless measurements of the magnetic field in a plasma can be realized by using the Zeeman splitting of spectral lines. The Zeeman components can be isolated by virtue of the differences in polarization. In practice this method is found to be much more difficult than the use of magnetic probes and is almost never used for the investigation of magnetic fields in plasmas, with the exception of cases of astrophysical interest, in which the method has been used for many years. (Cf., for example, the review in [25].) Certain questions associated with the resolution and direction of the Zeeman components of broadened spectral lines are considered in § 4.2.

In addition to investigating stationary, or quasistationary magnetic field distributions, magnetic probes are used in experiments on the observation and investigations of microfluctuations of magnetic fields in a plasma. For example, this problem arises in experiments on collisionless shock waves where microfluctuations can arise because of instabilities associated with Alfvén waves and magnetoacoustic waves in high-beta plasmas. Both instabilities result from a pressure anisotropy. The condition that leads to the excitation of the Alfvén instability is $p_{\parallel} > p_{\perp} + H^2/4\pi$. The ion-acoustic instability arises when $p_{\perp} > p_{\parallel}$. Here $p_{\parallel}$ and $p_{\perp}$ are the plasma pressures along the magnetic field and transverse to the magnetic field. The microfluctuations that arise as a result of these instabilities are characterized by a frequency $\Omega_H = eH/Mc$ and a spatial scale equal to the ion Larmor radius $\rho_i = Mvc/eH$. Using typical experimental values (the field frozen in the plasma $H \approx 200$ Oe and the directed flow velocity of the plasma $v = 3 \cdot 10^7$ cm/sec) the numerical values of these quantities are found to be $\Omega_H = 10^6$ sec^{-1} and $\rho_i = 10$ cm.

Low-inductance probes are used to obtain the spectrum of the microfluctuations (the number of turns is usually less than three or five); the signal from these probes is applied to the input of an oscilloscope without preliminary integration. The oscilloscope curve obtained in this way is analyzed in a Fourier series. In order to convert from the derivative to the spectral expansion of the magnetic field with respect to frequency, each term in the expansion must be divided by the corresponding frequency.

The spatial scale of the microfluctuations can be obtained by making correlation measurements with two probes.

The correlation coefficient for the readings of two probes separated by a distance Δx is determined by the expression

$$R = \frac{\int_0^T H(x_o) \cdot H(x_o + \Delta x)\, dt}{\sqrt{\int_0^T H^2(x_o)\, dt \cdot \int_0^T H^2(x_o + \Delta x)\, dt}}, \qquad (2.30)$$

where T is the lifetime of the process.

If the microfluctuations are of spontaneous nature the dependence of the correlation coefficients on the distance between probes is given by a Gaussian curve with a half-width equal to the spatial scale. If, however, the microflucutations represent some kind of wave process, the correlation curve will be of the nature of a damped oscillation, the distance between peaks corresponding to the wavelength.

Similar methods of analysis can be used with electric probes; however, the small space scale (order of the Debye radius) of the most common electrostatic oscillations in a plasma usually mean that it is not possible to obtain the dependence of the correlation coefficient on distance. Under typical laboratory conditions the dimensions of the probe are appreciably greater than the Debye radius. A much more convenient object for this kind of investigation is the interplanetary plasma, in which the Debye radius $\lambda_D \sim 10^3$ cm and is always much greater than the dimensions of the probes. In measuring microfluctuations of the magnetic field associated with pressure anisotropies frequently the inverse situation is encountered, that is to say, the Larmor radius of the ions in the interplanetary plasma (10^7 cm) is many orders of magnitude greater than the typical dimensions of a satellite. Hence, this investigation requires the availability of elements separated by distances of 10^5-10^6 cm; thus this measurement is much more complicated than the correlation measurement in a laboratory plasma.

The example given here shows that investigations of interplanetary and laboratory plasmas can be regarded from a single point of view; the type of measurement must be chosen to fit the problem at hand.

References

1. N. A. Kaptsov, Electronic Phenomena in Gases and in Vacuum [in Russian], Gostekhizdat, Moscow-Leningrad, 1947.

2. D. Bohm, The Characteristics of Electrical Discharges in a Magnetic Field, McGraw-Hill, New York, 1949.

3. F. F. Chen, Nucl. Energy, Part C, 7:47 (1965).

4. G. S. Janes and J. P. Dotson, Rev. Sci. Instrum., 35:1617 (1964).

5. E. E. Yushmanov, in: Plasma Physics and the Problem of a Controlled Thermonuclear Reaction, Pergamon, New York, 1958, Vol. 4.

6. E. O. Johnson and L. Malter, Phys. Rev., 80:158 (1950); I. A. Kovan et al., in: Plasma Diagnostics, Gosatomizdat, Moscow, 1963, p. 237.

7. S. Khvashchevskii, Probe Measurements of Plasmoids, Preprint of the Institute of Nuclear Physics in Warsaw (Sverk), 1963.

8. N. I. Ionov, Dokl. Akad. Nauk SSSR, 85:753 (1952).

9. I. M. Podgornyi and V. I. Sumarokov, Zh. Tekh. Fiz., 34:833 (1964) [Sov. Phys. — Tech. Phys., 9(5):635 (1964)].

10. N. I. Ionov, Zh. Tekh. Fiz., 34:769 (1965) [Sov. Phys. — Tech. Phys., 9(5):591 (1964)].

11. E. A. Lobikov and A. I. Nastyukha, Zh. Tekh. Fiz., 32:1223 (1962) [Sov. Phys. — Tech. Phys., 7(10):903 (1963)].

12. V. G. Averin, E. A. Lobikov, and A. I. Nastyukha, Zh. Tekh. Fiz., 34:1131 (1964) [Sov. Phys. — Tech. Phys., 9(6):879 (1964)].

13. Yu. G. Zubov, E. A. Koltypin, and E. A. Lobikov, Zh. Tekh. Fiz., 33:686 (1963) [Sov. Phys. — Tech. Phys., 8(6):513 (1963)].

14. J. H. Malberg, Rev. Sci. Instrum., 35:1622 (1964).

15. G. Ecker, W. Kroll, and O. Zoller, Ann. Physik, 10:222 (1962).

16. A. M. Andrianov et al., Progr. Nucl. Eng., 11:251 (1959).

17. L. A. Artsimovich et al., Atomnaya Énergiya, No. 3, 76 (1956).

18. V. S. Komel'kov, in: Proc. 2nd International Conference of the Peaceful Uses of Atomic Energy, Geneva, 1958.

19. L. S. Burgkhardt et al., in: Controlled Thermonuclear Fusion [in Russian], Atomizdat, Moscow, 1958, p. 30.

20. D. L. Tak, in: Proc. 2nd International Conference on the Peaceful Uses of Atomic Energy, Geneva, 1958.

21. A. M. Iskol'dskii et al., Zh. Eksp. Teor. Fiz., 47:774 (1964) [Sov. Phys. – JETP, 20(1):517 (1965)].

22. V. P. Smirnov and V. D. Rusanov, ZhETF Pis. Red., 2:356 (1965) [JETP Lett., 2(8):225 (1965)].

23. A. V. Bartov et al., Nucl. Fusion, Suppl. III, 1067 (1962).

24. Lovberg, in: Plasma Diagnostic Techniques, Academic Press, New York-London, 1966.

25. A. B. Severnyi, Usp. Fiz. Nauk, 88:3 (1966) [Sov. Phys. – Usp., 9(1):1 (1966)].

26. L. Biberman and B. Panin, Zh. Tekh. Fiz., 21:71 (1961).

Chapter 3

Determination of Electron Temperature
from Emission Intensity Ratios in Line Spectra

§ 3.1. Line Spectrum of a Plasma

In this chapter we wish to consider the emission intensity of the line spectrum of an atomic system located in a plasma. Atomic systems of this kind might be neutral atoms or ions which have not completely lost their electrons.

In order to understand the physical significance of the process responsible for the emission intensity of a line spectrum we first consider an idealized system that has a ground state and one excited state; for reasons of simplicity we neglect ionization and recombination processes. In what follows, the subscript "1" will denote the ground state while the subscript "2" will represent the excited state. If the density of atoms in the excited state is given by n_2, the energy of the line radiation at a frequency $\nu = (E_2 - E_1)/h$ per unit volume is given by

$$I = h\nu\, n_2\, A_{21}, \tag{3.1}$$

where A_{21} is the transition probability, which is usually called the Einstein coefficient:

$$A_{21} = \frac{g_1}{g_2} \cdot \frac{8\pi^2 e^2 \nu^2}{mc^3}\, f_{\text{abs}}, \tag{3.2}$$

where g_1 and g_2 are the statistical weights for the lower and upper states, respectively; e and m are the charge and mass of the elec-

tron; c is the velocity of light in vacuum; f_{abs} is the absorption oscillator strength. (Methods for computing f_{abs} are given in [1, 2]
and the numerical values for certain transitions are given in the
tables in [3].) The population in the excited levels is a complicated
function of the temperature and density and is quite sensitive to the
structure of the given atomic system, in particular, the existence
of metastable levels and cascade transitions from higher levels.
However, within the framework of the assumptions adopted here
the basic excitation processes are the absorption of a photon by the
atom and the collision of the atom with a free electron. Deexcitation occurs by emission from the excited atom or by collisions of
the second kind with an electron, in which case the excitation energy is transferred to the free electron. It is permissible to neglect collisions with ions if the ion temperature is not much higher
than the electron temperature. Under these conditions the equation
that describes the equilibrium between the excitation and deexcitation processes is

$$n_1 u(\nu) B_{12} + n_e n_1 \langle \sigma_1 v \rangle = n_e n_2 (\sigma_2 v) + n_2 A_{21} + n_2 u(\nu) B_{21}, \quad (3.3)$$

where n_1 is the density of atoms in the ground state, n_2 is the density of atoms in the excited state, n_e is the density of electrons, v
is the electron velocity, σ_1 is the effective cross section for excitation by electron impact, σ_2 is the effective cross section for
collisions of the second kind which quench excitation, $u(\nu)$ is the
spectral density of the radiation in the plasma, and B_{12} and B_{21}
are the Einstein coefficients for absorption and induced emission.
The symbol $\langle \ \rangle$ denotes an average over a Maxwellian distribution.

The plasma can be regarded as optically thin over a wide
range of values of density and temperature, that is to say, the absorption of radiation, including line emission, need not be taken
into account. The assumption that the radiation absorption in the
plasma can be neglected automatically means that terms containing
$u(\nu)$ are small, because the radiation escapes freely from the
plasma. In any given case the validity of this assumption can be
verified with respect to the intensities within multiplets. (The intensities of the lines in the multiplet, in accordance with the radiation rules, must be proportional to the quantity 2j + 1 for lines
which start from or terminate at a common level.)

Equation (3.3) is rather complicated even in the case of an optically thin plasma and, in general, is not convenient for the calculation of plasma parameters by measurements of spectral line intensities. In the two limiting cases of high density or low density the populations of the levels can be expressed by rather simple relations which can be used for the determination of the electron temperature. These limiting cases as well as the mutual effect of populations in different excited levels are analyzed in the following sections.

§ 3.2. Low-Density Plasma

We shall first consider conditions under which collisions of the second kind can be neglected. Physically this means that each excitation event by an electron collision is followed by deexcitation with the emission of a photon. This regime is characteristic of a low-density plasma and is observed in the solar corona under natural conditions. It is called the instantaneous emission regime of the corona regions. The second term is more popular but is less convenient since it is applied primarily to a distribution of multiply ionized ions; hence we will use the term "instantaneous emission." The density will be called low if excitation by collisions of the second kind can be neglected. In this case Eq. (3.3) assumes the form

$$n_1 n_e \langle \sigma_1 v \rangle = n_k \sum_{i=1}^{k-1} A_{ki}, \tag{3.4}$$

and consequently

$$I = h\nu n_e n_1 \langle \sigma_1 v \rangle \frac{A_{k1}}{\sum_{i=1}^{k-1} A_{ki}}. \tag{3.5}$$

Equations (3.4) and (3.5) are valid for an atomic system with many levels, in which case the summation is taken over all levels lying below the one being considered. The expressions are written under the assumption that the transitions from the higher levels to the level in question can be neglected. In passing we note that the absence of metastable levels and the low density of the plasma mean that we can assume with a high degree of accuracy that the density of atoms (or ions) in the ground state n_1 is equal to the density of the given atomic species (or ion species) n_i.

TABLE 3.1. Dimensionless Excitation Probability

$F(u)$	$\psi(x)$	Author
$\dfrac{3\,(u-1)}{u^2}$	$3\sqrt{x}\,e^{-x}\,[1-xe^x\,Ei\,(x)]$	Allen [3]
$\dfrac{0.45\,\ln u}{u-1}$	$0.225\,e^{-x}\,g\,(x)$	Knorr [5]
$\dfrac{2.72\,\ln u}{u}$	$2.72\sqrt{x}\,Ei\,(x)$	Post [6]
$\dfrac{\ln\,(\sqrt{u}+\sqrt{u-1}\,)}{u}$	$\dfrac{1}{2}\sqrt{x}\,e^{-\frac{x}{2}}\,K_0\left(\dfrac{x}{2}\right)$	Elwert [7]

The ratio $A/\Sigma\,A$ is equal to unity for the resonance line. If the quantities n_e and n_i are known on the basis of other measurements, and if the dependence of $\langle\sigma_1 v\rangle$ on temperature is known, the measurement of the intensity of the resonance line can be used to determine the electron temperature in the plasma. This method of determining the temperature is evidently one of the most approximate and least convenient since it requires supplementary measurements of n_e and n_i and also requires an absolute calibration of the intensity. The method becomes much simpler and the results become much more reliable if the measurement of the absolute intensity is replaced by a measurement of the relative intensities of two lines of the same atomic system. The intensity ratio for two lines is determined in the following way:

$$\frac{I_1}{I_2} = \frac{\nu_1\,\langle\sigma_1 v\rangle_1\,A_1\,\Sigma\,A_k^{(2)}}{\nu_2\,\langle\sigma_1 v\rangle_2\,A_2\,\Sigma\,A_k^{(1)}}\,. \tag{3.6}$$

The excitation cross sections can be computed from (3.7),

$$\sigma_1 = \frac{\pi e^4}{E^2}\,f_{\text{abs}}\,F\left(\frac{W}{E}\right). \tag{3.7}$$

The expression for the cross sections as averaged over a Maxwellian distribution is

$$\langle \sigma_1 v \rangle = \pi e^4 f_{abs} \sqrt{\frac{8}{\pi m E^3}} \, \psi \left(\frac{E}{kT} \right).$$

(3.8)

In Eqs. (3.7) and (3.8) W is the electron energy while E is the excitation energy. For the two-level model $E = E_2 - E_1$. The form of the function F(u) or ψ(x) depends on the method by which the cross sections are computed. The calculation of these functions can be carried out in the quasistatic approximation and the Born approximation as well as by other methods. The results of the calculation of F(u) and ψ(x) for dipole transitions are given in Table 3.1.

The following notation has been used in the table: g(x) = $x e^x$ Ei(x); $Ei(x) = \int\limits_{x}^{\infty} \frac{e^{-x}}{x} \, dx.$ A graphical representation of the function is given in [5].*

The behavior of the excitation cross section as a function of W/E is shown in dimensionless units in Fig. 3.1. A comparison of the functions ψ(E/kT) computed by various authors shows that the range is rather large so that these data can be used to determine temperatures from the intensity ratios of spectral lines.

In addition to considering the results of the calculations of the excitation cross sections contained in Table 3.1, it is also necessary to consider the work of Grizinskii [8]. The cross sections obtained by Grizinskii are not only convenient for the calculation of dipole transitions, but are also convenient for calculations in-

---

*In computing the function $Ei = \int\limits_{x}^{\infty} \frac{e^{-x}}{x} \, dx$ it is possible to make use of its series

representation

$$\int\limits_{x}^{\infty} \frac{e^{-x}}{x} \, dx = -0 \ 577 - \ln x + \frac{x}{1 \cdot 1!} - \frac{x^2}{2 \cdot 2!} + \frac{x^3}{3 \cdot 3!} - \cdots .$$

When

$$x \gg 1, \ Ei \approx \frac{e^{-x}}{x} \left(1 - \frac{1!}{x} + \frac{2!}{x^2} - \frac{3!}{x^3} + \cdots \right).$$

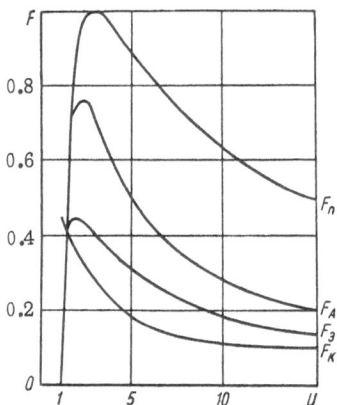

Fig. 3.1. Excitation cross sections computed by various methods.

volving transitions with a spin reorientation. One shortcoming of the expressions given in [8] is the fact they do not converge to the exact dependence ln W/W when W → ∞. This dependence is obtained for the ionization cross sections (cf. [4]) obtained by Drawin, which are also recommended [9] for use in the calculation of excitation probabilities.

At lower temperatures, where it is possible to work with lines due to neutral atoms, one can make use of the experimental data on the determination of the effective excitation cross sections as functions of the energy of an electron beam. It is still more convenient to use the experimental data for the excitation function for the lines. One advantage of the use of the experimentally obtained data for the use of the excitation functions for the lines rather than the levels is that the need for computing the transition probability is eliminated. Furthermore, in this case one not only takes account of collisional transitions from the ground state to this level (in the deexcitation of which the line being considered is distorted), but also the spontaneous transitions from higher levels. The most convenient atomic systems for the measurement of electron temperatures are neutral helium ions and helium-like ions. This feature is related to the appreciable difference in the form of the excitation function for singlet and triplet levels. The excitation of the triplet state occurs with the reorientation of a spin and in this case the excitation function is of a resonance nature, whereas the reduction in the probability for excitation of singlet levels with increasing energy is much smoother. Data for certain helium lines were first given in the work of Lees [10]. These data have been verified many times by various authors (cf. for example, [11, 12]). By averaging the excitation function over a Maxwellian distribution and substituting the results in Eq. (3.6) we can determine the electron temperature from the measured value of the intensity ratios.

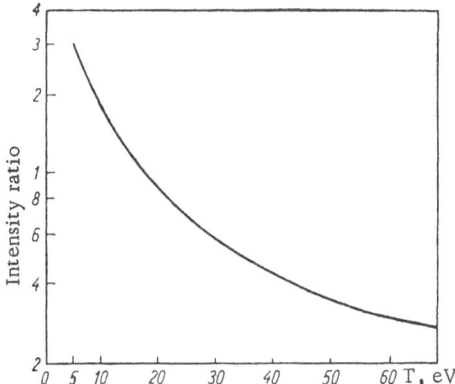

Fig. 3.2. The relative intensity of the helium
lines 4921 Å and 4713 Å as a function of electron
temperature.

A basic difficulty that is encountered in the measurement
of relative line intensities stems from the dependence of the sensitivity of the detector on the wavelength of the radiation. This
shortcoming is common to both photographic and photoelectric
detection methods. In the visible region of the spectrum the system can be calibrated by viewing a tungsten filament which is
maintained at a known temperature. The calibration of the detector should be carried out on the same spectral apparatus which
is used for measuring the temperature in order to avoid errors
associated with wavelength dependence of the optical absorption
in the device. It will be evident that the errors will be smaller
the closer the calibration wavelength to that of the spectral lines
being investigated. The dependence of the relative intensity for
two close-lying helium lines, 4921 and 4713 Å, as a function of
electron temperature is shown in Fig. 3.2 [13].

In determining the electron temperature by calculation or by
the experimentally obtained data for the excitation functions one
should keep in mind the possible errors associated with self-absorption; this is true not only for the lines whose intensities are
being measured, but also for lines corresponding to the transition
from the given excited state to the ground state. For example, in
the excitation of the helium 3^1P state, in addition to the 2^1S-3^1P
line at 5016 Å there is also a 1^1S-3^1P line at 537 Å which, in ab-

sorption, again excites the 3^1P state; consequently, for a given
temperature the intensity of the 5016 Å line will be greater than
would be expected from the experimental data obtained in excitation of isolated atoms by electrons. It has been shown in a recently published work [14] that the results obtained by Lees for
the 5016 Å line can only be used for a density of neutral helium
greater than $3 \cdot 10^{15}$ cm^{-3}, in which case self-absorption of the
537 Å line occurs. At a neutral gas density smaller than 10^{14}
cm^{-3} a plasma with characteristic dimensions of several centimeters is also optically transparent for the 537 Å line so that
the effective cross section for excitation of the 5016 Å line becomes several times smaller than would be expected from the
experimental data [10] on the determination of the excitation cross
section for the 5016 Å line.

There is one other process that imposes a strong limitation
on the usefulness of the helium method. In the analysis of the determination of electron temperature by the relative intensity of
heliums lines until very recently no account has been taken of collisions with electrons which lead to an equilization of the populations
of the levels corresponding to the same principal quantum number. There are dozens of published experimental papers in which
the helium method has been used for densities n ~ 10^{13}-10^{15} cm^{-3}
(for example, [15]). The results of measurements of the value of
T_e for these densities are found to be in contradiction with other
data, in particular those obtained in discharges at densities of
10^{14} cm^{-3} and electron temperatures of 10-20 eV, which are found
to contain a significant amount of neutral helium: according to the
available data on ionization cross sections at this temperature the
neutral helium should be burned out in ~10^{-6} sec. Calculations
carried out recently [16, 17] show that even at densities above 10^{12}
cm^{-3} collisions with electrons cause a strong modification of the
intensity of the helium lines and make the measurements essentially impossible. At higher densities, collisional transitions within a given principal quantum number lead to a relative depletion
of the 4^3S state which, in turn, leads to a change of the intensity
ratio for the 4713 and 4921 Å lines; frequently this effect is interpreted as an increase in electron temperature.

Experiments carried out with a plasma produced in a high-frequency discharge [18] have verified the validity of the theoretical
ideas concerning the applicability of the helium method for measuring temperatures at plasma densities greater than ~10^{12} cm^{-3}.

There is one other process that leads to a change in the relative intensities of the helium lines. As has been noted in [19, 20], the wave functions of the F levels in helium are essentially a superposition of approximately equal parts of singlet and triplet functions. Hence there is the possibility of transfer of excitation from the singlet states to the triplet states through the F level. In the establishment of a Boltzmann distribution over all the levels with principal quantum number equal to four the ratio of intensities of the 4713 and 4921 Å lines is approximately 0.3, which would correspond to an electron temperature higher than 50 eV, if σ were observed in the instantaneous emission regime, that is to say, if each excitation event due to electron impact were accompanied by a radiative transition. A typical example of this situation appears in the experimental conditions in [15]. In a cold plasma with a density of approximately 10^{15} cm^{-3} the electron temperature as measured from the intensity ratios of the helium lines at 4713 and 4921 Å was found to be 40 eV.

Experimental data on the excitation functions for the lines of neutral helium frequently cannot be used to measure T_e in a high-temperature plasma because even at a relatively low temperature the neutral helium atom may not exist in the plasma. In this case the measurement of T_e is possible only from a comparison of the lines of highly ionized atoms.

In this respect great interest attaches to helium-like and beryllium-like systems. Like helium these have terms corresponding to triplet and singlet structures; furthermore, they have the advantages compared with neutral helium that the levels with a given principal quantum number are shifted much more than in the case of helium so that the probability of collisional transitions between these levels is much smaller than for helium. Estimates show [17] that the limiting density for which it is still possible to measure the electron temperature in the helium-like ions CV, NVI, and OVII using the experimentally determined excitation function is approximately 10^{14} cm^{-3}.

The literature contains the rather pessimistic opinion [21] that the determination of the electron temperature from the intensity ratio for singlet and triplet lines in helium-like systems is not possible at densities beyond 10^{10} cm^{-3}. This conclusion is based on an erroneous estimate of the role of the metastable 2^3S state in the excitation of triplet levels. If the lifetime of the metastable state were determined only by the rate of metastable decay or the loss of ions to the wall (this effect is not very likely in a plasma confined by a magnetic field) at n $\sim 10^{10}$ cm^{-3} the rate of filling from higher triplet levels would exceed the rate of filling from the ground state and the dependence of the intensity of the triplet lines on density would be nonlinear. Actually, however, the basic process for the emptying of the metastable state is ion collisions with electrons [17]; at T$_e \sim 10$ eV, the cross section for this process is approximately $5 \cdot 10^{-16}$. This large ionization cross section has essentially very little effect on the metastable 2^3S state as far as population of the level with principal quantum number four is concerned at densities above 10^{11} cm^{-3}. Unfortunately, the application of the lines of helium-like ions is inconvenient in that bright lines of triplet and singlet levels are located in different spectral regions; consequently, the application of the lines of helium-like ions is inconvenient in that bright lines of triplet and singlet levels are located in different spectral regions; consequently, the determination of the intensity ratios for these lines is a rather complicated experimental problem. At the present time observations are made of singlet lines for transitions such as K^{1}P $\to 1^1$S and triplet lines for transitions such as 2^3P $\to 2^3$S, 3^3P $\to 2^3$S, and 3^3D $\to 3^3$P.

In this connection beryllium-like ions show a good deal of promise. These ions have pairs of intense singlet and triplet lines which are suitable for measurement and which lie in the visible region of the spectrum. Furthermore, the distribution of levels for the same principal quantum number in beryllium-like ions is larger than in helium-like ions. However, in contrast with helium-like ions, in beryllium-like systems the ionization potential for the 2^3P state is large so that the ionization cross section for ions in the metastable state is appreciably smaller than the cross section for formation due to excitation from the ground state; the basic mechanism for destroying the metastable 2^3P state is the intercombination transition 2^3P $\to 2^1$S. An estimate of the probabilities for the various processes shows that the excitation of triplet levels belonging to the same principal quantum number, three, occurs primarily from the ground state. Consequently, the triplet lines of beryllium-like ions 3^3L $\to 3^3$(L + 1) have excitation functions which differ sharply in form from the singlet lines 3^1L $\to 3^1$(L + 1). Both singlet and triplet lines of these transitions

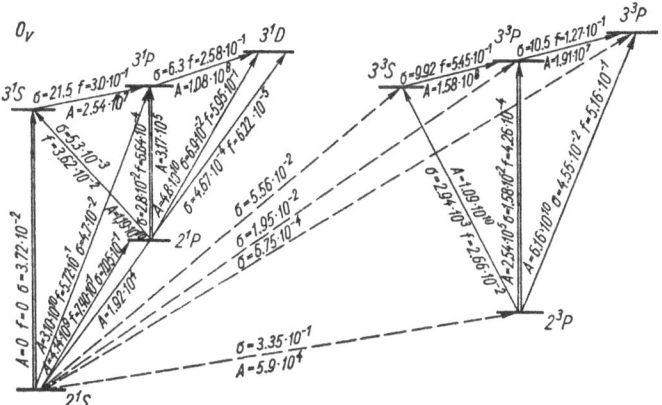

Fig. 3.3. Level diagram for the beryllium-like ion of oxygen.

of beryllium-like ions of carbon, nitrogen, and oxygen lie in the visible region of the spectrum and are convenient for experiment.

A level diagram for the beryllium-like ion of oxygen, with the probabilities for collisional and radiative transitions, is shown in Fig. 3.3. In Fig. 3.4 we show the critical values of the density for which collisional transitions prevent the measurement of electron temperature from the intensity ratio in beryllium-like ions.

There is no doubt that helium-like and beryllium-like ions will be widely used for the measurement of T_e when experimental data on the excitation functions become available. It would be highly useful to carry out a systematic investigation of the excitation functions for the lines of highly ionized atoms. It would be convenient to use for this purpose intense ion beams which could be bombarded by electrons.

In spite of the fact that present-day quantum mechanics does not allow us to compute excitation functions precisely, in the measurement of temperature it is frequently possible to use approximate data on the dependence of excitation cross sections on electron energy. For example, Kaufman and Williams [22, 23] have used the following expression for the intensity of the triplet line $2^3S - 2^3P$ in helium-like ions:

$$I = 2 \left(\frac{2kT_e}{\pi m} \right)^{1/2} n_i \cdot n_e \, Qh\nu \left(1 + \frac{E}{kT_e} \right) e^{-\frac{E}{kT_e}}.$$

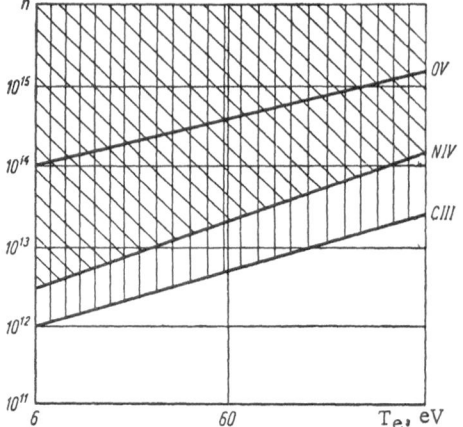

Fig. 3.4. Limiting concentrations for which it is
still possible to measure the electron temperature
using the lines of beryllium-like ions.

Here E is the energy of the 2^3P_2 level, Q is the excitation cross
section close to threshold, and n_i is the density of helium-like
ions. The following assumptions are used in the derivation of this
formula:

1) the excitation of the 2^3P_2 level occurs primarily from the
ground state and the cross section vanishes for electron energies
smaller than E and is constant for energies greater than E;

2) the deexcitation of the 2^3P_2 state occurs only as a result
of spontaneous transition with the emission of the $2^3S - 2^3P$ line.

Measurements of the electron temperature can be facilitated
if use is made of the lines from two helium-like ions since the
procedure does away with the need for carrying out absolute measurements in the determination of the electron temperature. Furthermore, the error due to the calculation of the cross section Q
has a much smaller effect on the result. An expression for the intensity ratio for the $2^3S - 2^3P$ lines for helium-like ions (on the
basis of the assumptions made above) is

$$\frac{I_1}{I_2} = \frac{n_1 Q_1 \nu_1 \left(1 + \dfrac{E_1}{kT_e}\right)}{n_2 Q_2 \nu_2 \left(1 + \dfrac{E_2}{kT_e}\right)} \exp\left(\frac{E_2 - E_1}{kT_e}\right). \tag{3.9}$$

In [23] the electron temperature in the Sceptre III device was determined by absolute and relative intensities of the BIII 2821.7 Å and CIV 2270.9 Å lines. The excitation cross sections for both lines were taken to be $0.2\pi a_0^2$, where a_0 is the radius of the first Bohr orbit. An admixture of carbon and boron was introduced into the discharge in measured amounts in the form of methane and dipropane. In substituting the numerical values in the formulas it was assumed that all of the atoms of carbon and boron in discharge were present in the form of helium-like ions; the measurements were carried out at the peak intensities for the 2821.7 and 2270.9 Å lines. The assumption concerning the small relative density of helium-like ions was verified by observation of the spatial distribution of the line emission for lines characteristic of different stages of ionization. It should be noted that the conditions in which helium-like ions predominate in the plasma are rather general since the potential of the helium-like ion is significantly greater than the ionization potential of the less easily ionized atoms. The question of ionization equilibria in the plasma will be discussed below. In returning to the results of Kaufman and Williams, it is interesting to note that in spite of the rather approximate nature of the excitation wavefunctions that have been chosen the values of the electron temperature determined by the absolute and relative intensities agree to within 30% ($T_e \sim 25$ eV).

As we have already noted, the principal difficulty in making a comparison of line intensities for lines in different regions of the spectrum lies in obtaining the spectral curve for the sensitivity of the measurement apparatus. In the visible region this difficulty can be overcome with relative ease by obtaining a calibration curve through the use of a tungsten filament lamp for which the spectral radiation characteristics are known. For work in the vacuum ultraviolet the calibration problem is much more complicated because of the lack of a reliable source which is calibrated over a wide region of the spectrum. Here the spectral device can be calibrated by another device, which acts as a monochromator in the desired spectral region. The intensity of the radiation beam selected by the monochromator is measured before striking the slit of the device being calibrated and at the exit. The transmission curves of the device for various diffraction gratings can differ significantly. The data available in the literature on the spectral sensitivity of photographic materials and sensitizers are very limited for the ultraviolet region. On the basis of the work reported in [24] it is usually assumed that the quantum yield of a sensitizer using sodium salicilate is independent of wavelength over a wide region of the vacuum ultraviolet.

§ 3.3. High-Density Plasma

We will regard a high-density plasma as one in which de-excitation is due to collisions of the second kind; thus, we are dealing with the case which, in some sense, is the opposite to the one considered in the preceding section. The condition that collisions of the second kind predominate can be written in the form

$$n_e \langle \sigma_2 v \rangle \gg A. \tag{3.10}$$

If this condition is satisfied, in the absence of photoexcitation an equilibrium is established between the inverse processes

$$n_1 \langle \sigma_1 v \rangle = n_2 \langle \sigma_2 v \rangle. \tag{3.11}$$

Using the well-known expression for the probability ratio for the forward and inverse processes

$$\frac{\langle \sigma_1 v \rangle}{\langle \sigma_2 v \rangle} = \frac{g_2}{g_1} e^{-\frac{E}{kT_e}} \tag{3.12}$$

(here E is the energy of the level; g_1 and g_2 are the statistical weights for the excited states and ground state, respectively), we obtain a Boltzmann population distribution over the levels. In this case the intensity of the spectral line corresponding to the transition from a state with excitation energy E to the ground state is given by

$$I = \frac{h \nu n_1 A g_2}{g_1} e^{-\frac{E}{kT_e}}. \tag{3.13}$$

The ratio of intensities for two lines for a high-density plasma (as in Eq. (3.6), for a low-density plasma) is independent of density,

$$\frac{I_1}{I_2} = \frac{\nu_1 g_1 A_1}{\nu_2 g_2 A_2} e^{-\frac{E_1 - E_2}{kT}}. \tag{3.14}$$

In this expression the subscripts 1 and 2 denote two different excited levels of the same atom with excitation energies E_1 and E_2, respectively. The transitions from the levels 1 and 2 lead to the emission of lines with intensities I_1 and I_2 and these can be transitions either to the ground state or to other excited states. We recall that the notation is different from that used in all expressions written for the two-level model (§3.2): in that case the subscript 1 denotes the ground state while the subscript 2 denotes the single excited state.

Thus, if the basic process for deexcitation is a collision of the second kind, that is to say, if the level population is described by the Boltzmann relation, then Eq. (3.14) is valid for any pair of lines. Consequently, knowing the transition probability and measuring the intensity ratio of the lines we can determine the temperature from the relation

$$T_e = \frac{\Delta E}{k} \cdot \frac{1}{\ln\left(\frac{\nu_1 g_1 A_1 I_2}{\nu_2 g_2 A_2 I_1}\right)}. \tag{3.15}$$

We now wish to consider in somewhat greater detail the region of applicability of this method of determining electron temperature. Using the expression for excitation probability and the principle of detailed balance we can obtain the probability for collisions of the second kind that lead to excitation,

$$\langle \sigma_2 v \rangle = \pi_e^4 f_{\text{abs}} \sqrt{\frac{8}{\pi m E^3}} \psi\left(\frac{E}{kT_e}\right) \frac{g_1}{g_2} e^{\frac{E}{kT_e}}. \tag{3.16}$$

Substituting this expression and the expression for the probability of a radiative dipole transition in Eq. (3.10) we obtain the criterion for the establishment of a Boltzmann population distribution,

$$e^2 h^2 c^3 \sqrt{\frac{m}{8\pi^3 E^7}} e^{\frac{E}{kT_e}} \psi\left(\frac{E}{kT}\right) > 1. \tag{3.17}$$

For purposes of making rough estimates with $E \sim kT$ we use the expression

$$10^{-14}\frac{n_e}{E^{7/2}(\text{eV})} > 1. \tag{3.18}$$

Substitution of numerical values in Eq. (3.18) shows that the establishment of a Boltzmann population distribution in the levels is rather infrequent. A Boltzmann distribution obtains in a plasma only at very high temperatures, for example, in a condensed spark [25].

For levels close to the ground state the Boltzmann population distribution is also observed at the time of maximum compression of an intense pulsed discharge [26, 27]. Although there is a rapid ionization in this case, with the production of more ionized states, the Boltzmann distribution can be established, since the required time is much smaller than the time required for establishing an ionization equilibrium.

Even if (3.18) is satisfied this does not necessarily mean that the population of all levels in the atom is described by the Boltzmann formula, since the lifetime of an electron in higher levels is generally determined by ionization rather than collisions of the second kind. The cross section for ionization for higher excited states can be several orders of magnitude greater than the cross section for excitation from the ground state. As a result, ionization of the higher excited states occurs before collisions of the second kind; consequently, the population of the higher levels is determined primarily by the equilibrium with the electron continuum.

In considering thermal equilibrium between the atomic levels and the continuum Wilson has applied another criterion for the establishment of a Boltzmann equilibrium [28]. This criterion is of the form

$$n > 10^{13} E_i T_e^{1/2}. \tag{3.19}$$

Here E_i is the ionization energy of an atom in a given excited state in eV.

If partial trapping of the radiation occurs, the conditions for establishment of a Boltzmann population distribution are facilitated and the criterion in (3.18) becomes less stringent; however, the measured intensities of the spectral lines in the presence of self-absorption do not correspond to the total emission of all atoms.

Consequently these intensities become inconvenient for the measurement of electron temperature. The presence of self-absorption can be easily checked by varying the relative intensities of lines in a multiplet.

Most frequently the electron density of a plasma is such that the contributions due to spontaneous emission and collisions of the second kind are comparable. In this case the line intensity is given by

$$I = h\nu \frac{g_2}{g_1} e^{-\frac{E}{kT_e}} \left(1 + \frac{A}{\langle \sigma_2 v \rangle n_e} \right). \tag{3.20}$$

Unfortunately this expression can only be used when the transitions between the given excited state and states other than the ground state can be neglected.

In the analysis of population levels of atomic systems presented in this and the preceding sections, it follows that there are two limiting cases (high and low concentrations) in which the ratio of intensities of spectral lines depend only on the electron temperature. As a rule neither of these cases is realized in a laboratory plasma and the ratio of the line intensities is a function of temperature as well as concentration. A precise knowledge of the probabilities for all radiation transitions could be used to determine the electron temperature and the density by measurements of the intensity ratios of three different lines from the same atom (or ion). In this measurement we have two independent equations to determine two unknowns:

$$I_1/I_3 = f_1(n, T_e) \qquad I_2/I_3 = f_2(n, T_e).$$

A basic difficulty in the use of the three-line method lies in the choice of similar lines; that is to say, it is not always possible to make quantitative calculations if one takes account of all of the experimental processes and their effects on the population of the appropriate levels. The first attempt to use the three-line method to measure the parameters of the high-temperature plasma was that of Suckewer [36, 37].

An analysis of the intensity ratio shows that it is easier to find lines which are suitable for determining density than for measurement of electron temperature. The intensity ratio for lines

having close-lying upper levels [so that $(E_1 - E_2)/kT_e \ll 1$] depends on the plasma density and is essentially independent of temperature under appropriate conditions. Typical examples are the
helium lines that have been considered in the preceding section,
where the intensity ratio for densities higher than 10^{12} cm^{-3} is
essentially independent of temperature. A detailed analysis carried out by Suckewer has shown that in determining densities in
the range 10^{13}-10^{17} cm^{-3} and $T_e \sim 10$ eV it is possible to use the following pairs of spectral lines in ions: CIV, 384 Å and 312 Å, NV,
247 Å and 209 Å, and OVI, 173 Å and 150 Å. Curves showing the
dependence of the intensity ratios for these lines are given in [37].

§ 3.4.   Ionization State of a Plasma

In determining the electron temperature from the relative
intensity of spectral lines from atoms or ions in various ionization
states it is necessary to know the relative densities of these ions.
These densities are not only functions of the electron temperature,
but also depend on a number of other factors that determine the
nature of the ionization equilibrium that obtains under given conditions. The equation that describes the ionization equilibrium can
be written

$$n_i n_e S_{i \to i+1} + n_i Q_{i \to i+1} = n_{i+1} n_e^2 S_{i+1 \to i} + n_{i+1} n_e Q_{i+1 \to i} \quad (3.21)$$

where 1) $S_{i \to i+1} = \langle \sigma_i v \rangle$ is the probability of ionization by electron impact; 2) $Q_{i \to i+1}$ is the probability of photoionization; 3)
$S_{i+1 \to i}$ is the probability of three-body recombination; 4) $Q_{i+1 \to i}$
is the probability of radiative recombination in a binary collision.

In low-density plasmas, in which the level populations are
not described by the Boltzmann distribution, ionization occurs primarily in the ground state.

The processes listed above do not exhaust all the possibilities
for ionization and recombination. In particular, no account has
been taken of ionization due to ion−ion collisions and collisions
which lead to the loss of more than one electron. In all cases of
ionization equilibrium that are of practical interest only electron
collisions are important. Processes 1-3 and 2-4 are reversible,
that is to say, in the state of thermodynamic equilibrium a balance

obtains between each pair of reversible processes. [Strictly speaking, in addition to the photorecombination term there should be a term corresponding to induced photorecombination on the right side of Eq. (3.21). However, the contribution of this term is almost always small and it will be neglected.]

At thermodynamic equilibrium, i.e., when the probabilities of the reversible processes are equal (cf., for example, [29]) the relative densities of ions in two successive stages of ionization are described by the Saha equation,

$$\frac{n_{i+1} \, n_e}{n_i} = 2 \frac{u_{i+1}}{u_i} \cdot \frac{(2\pi m k T)^{3/2}}{h^3} \, e^{-\frac{E_i}{kT}}, \tag{3.22}$$

where n_i is the density of ions with charge ie; E_i is the ionization potential of the ion, that is to say, the energy necessary for the transition from the i to the i + 1 ionization state; h is Planck's constant; $u_i = \sum g_k e^{-\frac{E_k}{kT}}$ is the statistical sum over all the excited levels of the ion.

In the derivation of the Saha equation the ionized atom is taken together with the free electrons as a single entity having a positive energy; it is further assumed that the population of levels having positive energies is described by the Boltzmann distribution. The Saha equation applies only in plasmas of rather high density. Even in an intense pulsed discharge in the compressed state with a plasma density of $\sim 10^{17}$ cm^{-3} we find that the Saha equation cannot be used. It appears that the only case that can be described by the Saha equation in a gas discharge is the plasma in a condensed spark [23] (n = 10^{19} cm^{-3}). The relative density of different ions of the same element as a function of the temperature for an electron density n = 10^{17} cm^{-3} have been computed under the assumption of equilibrium between reversible processes; the results are shown in Fig. 3.5.

It is much more common to encounter a different kind of ionization equilibrium; this equilibrium was first described in astrophysics, and is known as the Elwert equilibrium [30]. This equilibrium obtains if the following two conditions are satisfied:

1. The contribution due to photoionization is small compared with the ionization due to electron impact. This condition is almost

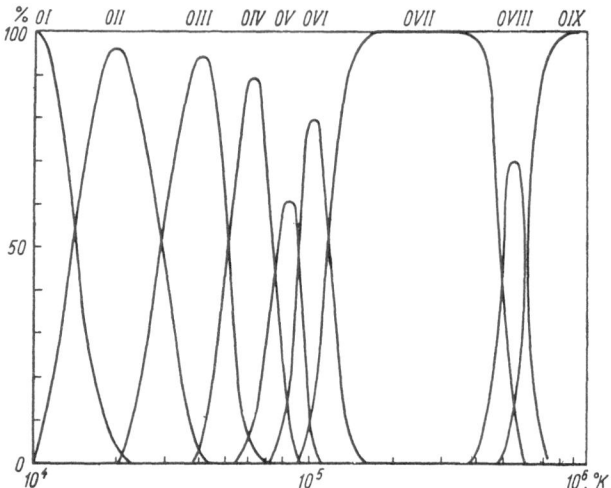

Fig. 3.5. The concentration of ionized atoms as a function of
temperature for $n = 10^{17}$ cm^{-3} computed under the assumption
of a Saha equilibrium. Actually, a high density is required to
establish a Saha equilibrium.

always well-satisfied because even in a plasma with a high optical
density only the radiation of the resonance lines is trapped.

2. Recombination occurs through binary collisions. The contribution of three-body collisions in the recombination process is
negligibly small.

The Elwert equilibrium is realized at low plasma densities
and describes the ionization state in the solar corona and is frequently called the corona distribution or the corona equilibrium.
The expression for the corona equilibrium can easily be obtained
if the ionization and photorecombination probabilities are known.
One requirement for this equilibrium is that three-body recombination must play a small role compared with photorecombination; hence, in understanding the limits of applicability of the
corona distribution we must also consider the probability of three-
body recombination. We shall now consider briefly the results of
calculations for these various probabilities.

1. Ionization by Electron Impact. The differ-
ential cross sections for ionization and the cross sections averaged

TABLE 3.2.

Author	$F_i(u)$	$\psi_i(x)$
Thomson [30]	$\dfrac{4(u-1)}{u^2}$	$4\sqrt{x}\,e^{-x}K(x)$
Elwert [30]	$\dfrac{2(u-1)}{u^2}[1 +0.3(u-1)]$	$\dfrac{1}{2}\sqrt{x}\,e^{-x}K(x)\{1 +0.3[(xK(x))^{-1}-1]\}$
Bethe [30]	$2.18\dfrac{\ln u}{u}$	$2.18\sqrt{x}\,E_i(x)$
De la Ripelle [5]	$\dfrac{2.86\ln u}{u+1}$	$2.86\sqrt{x}\,E_i(x)[1-\dfrac{1}{2}K(x)]$
Drawin [31]	$2.66\dfrac{u-1}{u^2}\ln(1.25\,u)$	$2.66\left\{\ln 1.25\,e^{-x}K(x)+Ei(x)\right.$ $\left. -x\dfrac{2-x}{5.42}\right\}$

over a Maxwellian distribution are expressed by formulas similar to those which have been obtained above for the excitation cross sections,

$$\sigma_i = \frac{\pi e^4}{E_{ik}^2}\,f_1\,\xi_k\,F_i\left(\frac{W}{E_{ik}}\right);\tag{3.23}$$

$$\langle\sigma_i v\rangle = \pi e^4\,\xi_k\,\sqrt{\frac{8}{\pi m\,E_{ik}^3}}\,\psi_i\left(\frac{E_{ik}}{kT}\right).\tag{3.24}$$

In these expressions W is the electron energy; E_{ik} is the ionization energy for a level with principal quantum number k; ξ_k is the number of electrons in this level; f_1 is a correction factor which is approximately equal to unity. Expressions for the functions $F_i(u)$ and $\psi_i(x)$ in various approximations are given in Table 3.2.

In Table 3.2, for reasons of simplicity we have introduced the notation

$$K(x) = 1 - xe^x\,Ei(x).$$

The best agreement with experiment is obtained using the ionization cross sections computed by Drawin. We note in passing that the dimensionless factor F_i computed by Drawin is also recommended in calculations of the excitation cross sections.

2. Probability of Three-Body Recombination. Three-body recombination is the name given to the elementary interaction event between an ion and two electrons in which one of the electrons is captured in an orbit of the ionized atom and excess energy is transferred to the other free electron. Three-body recombination is the inverse of ionization by electron impact. The ratio of probabilities for these processes is determined by the ratio of densities of ions with charge different from unity under conditions of thermodynamic equilibrium. Making use of the Saha formula and the expression for the ionization probability, that is to say, the process which is the inverse of three-body recombination, we have

$$S_{i+1 \to i} = \langle \sigma_i v \rangle \frac{u_i}{u_{i+1}} \cdot \frac{h^3}{(2\pi mkT)^{3/2}} e^{\frac{E_i}{kT}}. \tag{3.25}$$

3. Probability for Photorecombination. This process is the inverse of photoionization. From the expression for the probability for photoionization and the principle of detailed balance we can easily obtain the probability for photorecombination. The following expression for the probability of photorecombination for hydrogen-like ions is given by Seaton [32]:

$$Q_{i+1 \to i} = 5.2 \cdot 10^{-14} z \sqrt{\frac{E_i}{kT}} M\left(\frac{E_i}{kT}\right), \tag{3.26}$$

where M is a function that is tabulated in [32]. Equation (3.26) can also be used for making rough estimates of recombination in non-hydrogen-like atoms, but in this case it is preferable to use the data given by Seaton and Burgess [33].

Comparing the probabilities for three-body recombination and photorecombination we can easily obtain a critical value for the density at which the probabilities of these processes become comparable. According to the data given by Drawin in a hydrogen plasma the probabilities for these two modes of recombination become the same at a density $\sim 5 \cdot 10^{17}$ cm^{-3}. For purposes of rough estimates the critical density can be given by the expression

$$n_{\mathrm{cr}} \sim 10^{13} E_i^2 T^{3/2}. \tag{3.27}$$

The electron temperature and ionization energy are expressed in electron volts. A somewhat different critical density, i.e., a density above which the corona equilibrium no longer applies, is defined by Wilson. Here, n_{cr} is defined from the motion of the thermal limit. The population of levels lying above the thermal limit is determined by the equilibrium with the continuum so that excitation to these levels can be regarded as ionization. The formula that relates n_{cr} with the temperature and the ionization potential in this analysis is

$$n_{\mathrm{cr}} = 10^{11} E_i^{3/2} T_e^2. \tag{3.28}$$

At densities such that $n_e < n_{\mathrm{cr}}$, that is to say, under conditions of low radiation density, the corona equilibrium must be established in the plasma in the course of a definite time period. In this case the ratio of the density of the (i + 1) multiply ionized atom to the density of the i multiply ionized atom is [30]:

$$\frac{n_{i+1}}{n_i} = 8.3 \cdot 10^5 \frac{\xi_k}{g_k k_i} \left[\frac{E_i^{(H)}}{E_i} \right]^2 \frac{kT_e}{E_i} e^{-\frac{E_i}{kT_e}}, \tag{3.29}$$

where ξ_k is the number of electrons in the outer shell of the i ionized ion; k_i is the principal quantum number of the ground state of this ion; g_k is a factor that takes account of recombination to the upper level; the value of this factor is given by $1.4 < g_k < 4$; $E_i^H = 13.6$ eV.

Equation (3.29) describes the corona equilibrium and is convenient in that the ratio of densities for a given atomic species is a function of the temperature only. The exponential nature of the dependence of the ion density on ionization potential for a fixed value of the temperature means that only a limited number of ions can exist for a given temperature. The ionization potential of the ions whose species predominates is (5-10)kT. Hence the form of the spectrum itself can be used to make a fairly accurate estimate of electron temperature in the plasma. In Fig. 3.6 we show the relative densities of ions of different multiplicity as a function of temperature in an Elwert equilibrium. It is evident from the figure that lines of highly ionized states dominate even at temperatures of tens of electron volts. Almost all of the lines of these

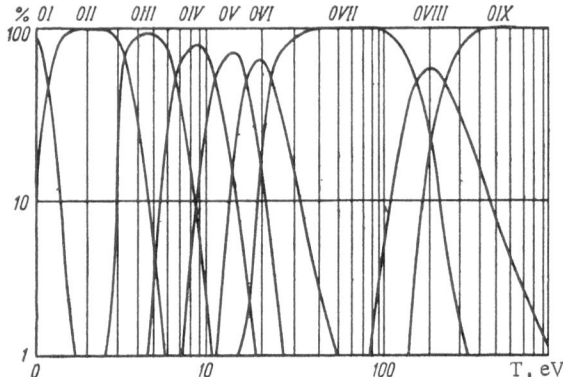

Fig. 3.6. The concentration of oxygen ions as a function
of temperature in the corona equilibrium.

states are found in the vacuum ultraviolet. In Fig. 3.7 we show
typical spectra for impurities in a hydrogen discharge at T_e =
25 eV. In the spectrogram the brightest lines are those of ions
which exist in the plasma at temperatures of 20-30 eV under conditions of the Elwert equilibrium. At higher temperatures all of
the important lines are found in the soft x-ray region. The spectra shown in Fig. 3.7 are obtained at a plasma density of 10^{17} cm^{-3}.
It is interesting to note that if a Saha equilibrium were established
in the plasma at this density (cf. Fig. 3.5) the nature of the spectrum would be entirely different (in particular, the OV and OVI
lines would not be observed).

Very frequently the emission spectrum of a plasma contains
lines due to weakly ionized atoms and neutral atoms in addition to
lines due to highly ionized atoms. The presence of lines due to
neutral or weakly ionized atoms indicates a lack of ionization
equilibrium or shows that the apparatus records radiation that
comes from both the hot and cool regions of the plasma.

The time required to establish equilibrium is very important
in research on controlled thermonuclear reactions where, as a rule,
the plasma exists for fractions of a second. In a paper by Mc-
Whirter [34], a criterion is given for the establishment of equilibrium, the criterion being obtained on the basis of numerical
values for the ionization and recombination probabilities given in
the paper. The criterion is of the form

$$nt > 10^{12} - 10^{13} \tag{3.30}$$

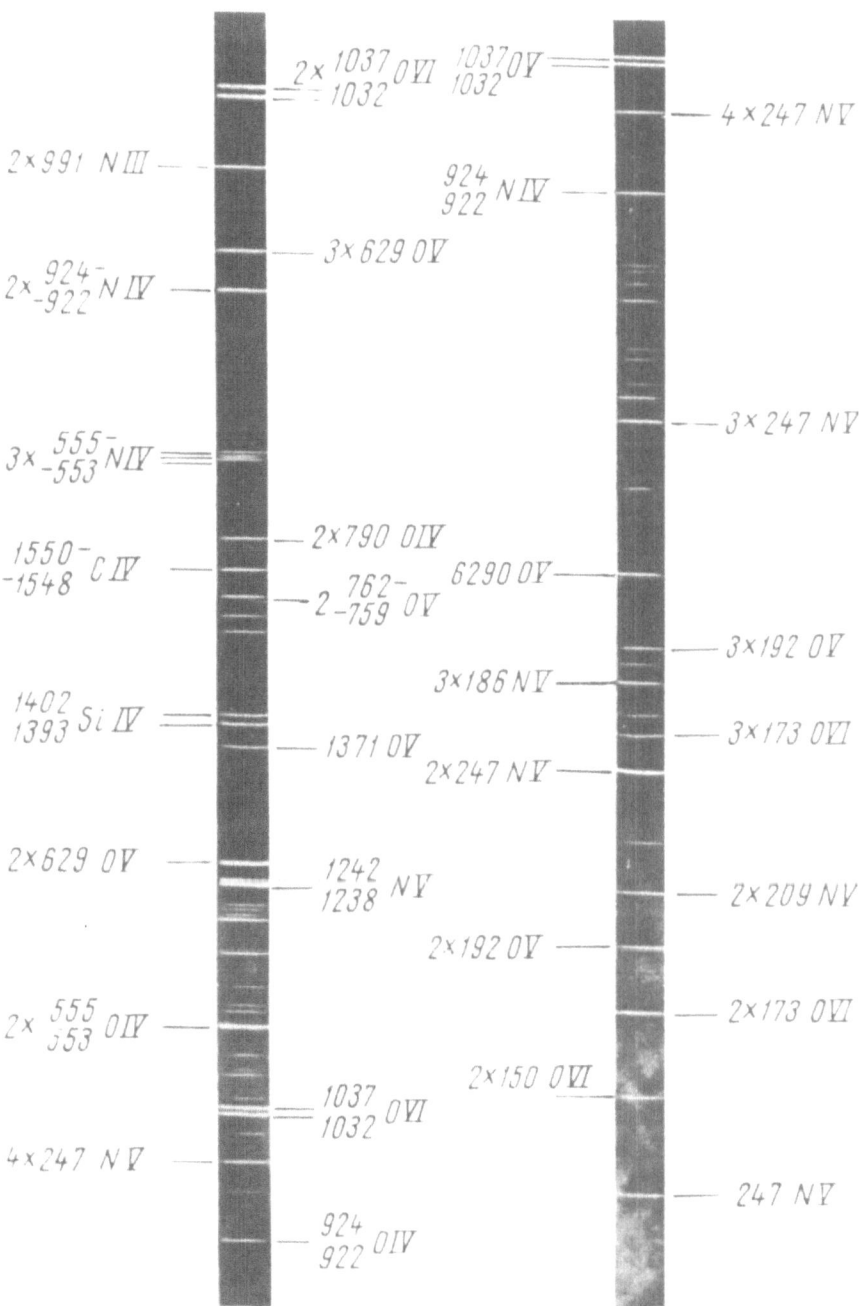

Fig. 3.7. The emission spectrum from the plasma in a pulsed discharge.

TABLE 3.3. Equilibrium Ionization Relaxation Time

Device	Plasma lifetime, sec	Density, cm⁻³	Equilibrium ionization relaxation time, sec
Pulsed discharge	10^{-6} (compressed state)	10^{17}-10^{18}	10^{-4}-10^{-6}
Circular discharge stabilized by magnetic field	10^{-3}	10^{13}-10^{14}	10^{-1}-10^{-2}

and determines the time necessary for the formation of the entire chain of ionization states.

It will be evident that the time for establishing ionization equilibrium in the corona is such that an equilibrium of this kind is rarely established in a plasma in a thermonuclear experiment aside from those in which the condition for existence of equilibrium is known to be satisfied beforehand (3.28).

In Table 3.3 we present data for the time required to establish ionization equilibrium in several typical thermonuclear devices.

If the condition $nt > 10^{13}$ cm⁻³ is not satisfied, using the data for ionization probabilities at various temperatures and the known electron density, it is possible to estimate the electron temperature from the time at which a given ion vanishes. Curves of the reciprocal ionization probability $\tau = \frac{1}{n < \sigma_i v >}$ for various ions of oxygen at a density $n = 10^{15}$ cm⁻³ are given in Fig. 3.8. In making these estimates it is necessary to have some assurance that the lifetime of the ion is determined by the ionization time rather than other processes (for example, charge exchange or diffusion).

We now wish to consider a characteristic example of an erroneous interpretation of the results of an observation of the time behavior of the intensity for the lines due to atoms and ions. This example is interesting in that the method is based on the analysis of a properly executed experiment. However, a small, and at first glance, trivial change in the conditions makes the analysis ambigu-

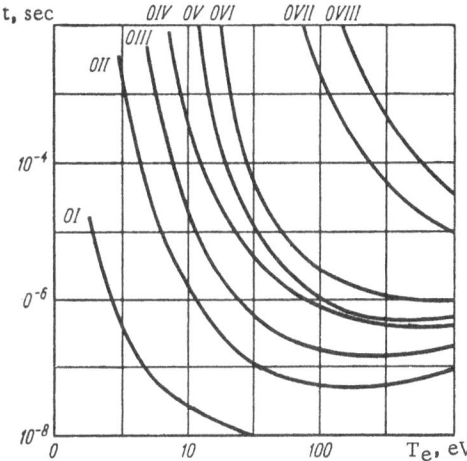

Fig. 3.8. The ionization time for oxygen at a density 10^{15} cm^{-3}.

ous. In [35], in determining the electron temperature of a rapidly heated plasma the rate of reduction of the intensity of the spectral lines for He I and He II was measured. Since the hydrogen plasma was confined by a magnetic field and the energy losses due to excitation and ionization of helium can be neglected (the helium was present in an amount less than 1%) the decay of intensity of the lines in this case actually characterizes the burnout time. Using the measured burnout time for He I and He II and the known value for the electron density it was possible to estimate the electron temperature. Under these conditions the method is free from objections. Unfortunately, this same method of tracking has been used incorrectly by a number of authors. For example, simpler measurements were carried out using the helium impurity in the amount of 50% and referring to the work of [35] the burnout time for neutral and singly ionized helium under conditions of rapid heating were determined from the decay in the intensity of the He I and He II lines. Actually, the intensity decay under these conditions is undoubtedly due to the reduction of electron temperature caused by energy losses incurred in the excitation and ionization of helium atoms. In other words, this is a typical situation in which unique analysis of the results of the measurements is not available; in this case it is easy to be led into error by carrying out an analysis which provides the desired results.

References

1. D. Bates and A. Damgaard, Phys. Trans., A242:101 (1949).
2. I. B. Levinson and A. A. Nikitin, Handbook on the Theoretical Computation of Line Intensities in Atomic Spectra [in Russian], Izd-nie LGU, Leningrad, 1962.
3. C. W. Allen, Astrophysical Quantities, Oxford Univ. Press, 1960.
4. A. V. Vasil'ev, G. G. Dolgov-Savel'ev, and V. I. Kogan, Nucl. Fusion, Suppl., Part 2, p. 655 (1962).
5. G. Knorr, Z. Naturforsch., 13a:941 (1958).
6. R. Post, International Summer Course in Plasma Physics, Riso (Denmark), Report No. 18 (1960).
7. G. Elwert, Z. Naturforsch., 9a:637 (1958).
8. M. Gryzinsky, Phys. Rev., 115:374 (1959).
9. I. I. Sobel'man, Introduction to the Theory of Atomic Spectra [in Russian], Fizmatgiz, Moscow (1963).
10. J. H. Lees, Proc. Roy. Soc., 173:1372 (1932).
11. S. É. Frish, Optical Spectra of Atoms [in Russian], Fizmatgiz, Moscow-Leningrad, 1963.
12. R. M. St. John, et al., Phys. Rev., 134A:888 (1964).
13. P. S. Thoneman, in: Dickerman Optical Spectrometric Measurements of High Temperature, Univ. Chicago Press, 1961, p. 56.
14. R. M. St. John, et al., J. Opt. Soc. America, 50:28 (1960).
15. V. J. Masahiro, Phys. Soc. Japan, 18:419 (1963).
16. I. M. Podgornyi and G. V. Sholin, Dokl. Akad. Nauk SSSR, 160:575 (1965) [Sov. Phys. — Dokl., 10(1):48 (1965)].
17. I. M. Podgornyi and G. V. Sholin, 7th International Conference on Phenomena in Ionized Gases, Belgrade, 1965.
18. V. L. Vdovin, I. M. Podgornyi, and V. D. Rusanov, Atomnaya Énergiya, 20:148 (1966) [Sov. Atomic Energy, 20(2):180 (1966)].
19. R. M. St. John and R. G. Fowler, Phys. Rev., 122:1813 (1961).
20. C. C. Lin and R. G. Fowler, Ann. Phys., 15:461 (1961).
21. L. Heroux, Proc. Phys. Soc., 83:121 (1964).
22. S. Kaufman and R. V. Williams, Nature, 182:557 (1958).
23. R. V. Williams and S. Kaufman, Proc. Phys. Soc., 75:329 (1960).
24. F. S. Johnson, et al., J. Opt. Soc. America, 41:702 (1951).
25. S. L. Mandel'shtam and N. I. Sukhodrev, Zh. Eksp. Teor. Fiz., 24:701 (1953).
26. V. D. Pis'mennyi, I. M. Podgornyi, and Sh. Sukever, Zh. Eksp. Teor. Fiz., 43:2008 (1962) [Sov. Phys. — JETP, 16(6):1416 (1963)].
27. Yu. K. Zemtsov, V. D. Pis'mennyi, and I. M. Podgornyi, Dokl. Akad. Nauk SSSR, 155:312 (1964) [Sov. Phys. — Dokl., 9(3):219 (1964)].
28. R. J. Wilson, Quant. Spectr. Rad. Transf., 2:477 (1962).
29. A. Unsold, Physics of Stellar Atmospheres [Russian translation], Izd. Inostr. Lit., Moscow, 1949.
30. G. Elwert, Z. Naturforsch., 7a:432,703 (1952).
31. H. W. Drawin, Z. Phys., 164:513 (1961).
32. M. J. Seaton, Mon. Not. Roy. Astr. Soc., 119:81 (1959).

33. M. J. Seaton and A. Burgess, Mon. Not. Astr. Soc., 120:121 (1959).
34. P. W. McWhirter, P. Proc. Phys. Soc., 75:520 (1960).
35. M. V. Babykin et al., Zh. Eksp. Teor. Fiz., 46:511 (1964) [Sov. Phys. — JETP, 19(2):349 (1964)].
36. S. Suckewer, Phys. Lett., 24A:284 (1967).
37. S. Suckewer, Phys. Rev., 170:239 (1968).

Chapter 4

Determination of Plasma Parameters
from the Shape of Spectral Lines

§ 4.1.   Introduction

The line shapes of spectral lines of atoms or ions in a plasma
are distorted as a consequence of various processes, each of which
leaves its own mark on the shape of the line. Hence, in spite of
the fact that the problem of the line shape in general form leads
to insurmountable mathematical difficulties, an analysis of the
isolated sections of a given line can sometimes be used to obtain
data on plasma parameters that do not affect each other. For example, if the broadening of the line is due simultaneously to the
Doppler effect and the Stark effect (in the statistical field of the
ions) the central part of the line contains information on the ion
temperature while the line shape in the wings contains information on the density of charged particles. It is also sometimes
possible to separate different broadening factors by exploiting the
difference in polarization in different components of the line. However, the greatest reliability is found in the case in which the line
shape is determined completely by one effect and the distortion
due to the spectral device and the detection method can be neglected.

If it is possible to make accurate controlled experiments, and
if careful comparisons of the theoretical and experimental line
shapes show that the contribution to the broadening due to other
factors is insignificant, it is not necessary to carry out a complicated comparison between the theoretical and experimental data

in every measurement. A simple measurement of the line width at the half-height points or, as it is usually called, the half-width, becomes adequate. Then, using some simple relations, it is possible to determine that plasma parameter which is responsible for the line width under a given set of experimental conditions.

In the following sections of this chapter we shall consider the various factors that lead to broadening of spectral lines and discuss the possibility of using these factors in plasma diagnostics.

§ 4.2. Determination of Ion Temperature

from the Doppler Broadening

of Spectral Lines

The temperature of neutral atoms or incompletely ionized atoms can be easily determined from the Doppler broadening of the spectral lines emitted by these systems if other broadening mechanisms are not important under a given set of experimental conditions. Neglecting the natural width of the spectral lines, we see that the emission spectrum of an atom will show a monochromatic line with wavelength λ_0. If the emission source moves toward the spectral device then this device will record a Doppler reduction in the wavelength given by $\Delta\lambda = \lambda_0(v_x/c)$, where v_x is the projection of the velocity along an axis passing through the emission source and the input aperture of the spectral device. The radiation from a source moving in the direction of the spectral device has a wavelength given by

$$\lambda = \lambda_0\left(1 + \frac{v_x}{c}\right). \tag{4.1}$$

The spectral location of the radiation of a large number of atoms whose directed motion is randomly distributed in space does not give a monochromatic line, but rather a range of wavelengths close to the value λ_0. The emitted energy in an interval $d\lambda$ in the vicinity of λ is determined by the density of atoms having the velocity component v_x in the vicinity of the value given by Eq. (4.1).

If the atoms exhibits a Maxwellian distribution over velocity, the probability that the velocity component will have a value lying in the limits v_x, $v_x + dv_x$ is given by the familiar expression

$$\frac{dn}{n} = \left(\frac{M}{2\pi k T_i}\right)^{1/2} e^{-\frac{Mv_x^2}{2kT}} i dv_x, \tag{4.2}$$

where T_i is the ion temperature.

Using Eqs. (4.1) and (4.2) we can easily obtain the distribution of spectral density within the given line,

$$I = I_0 e^{-\frac{Mc^2}{2kT_i}\left(\frac{\Delta\lambda}{\lambda_0}\right)^2} \tag{4.3}$$

In §4.1 we have noted that there are a number of other factors that can cause broadening of spectral lines in addition to the Doppler effect. Effects that depend on the plasma parameters will be analyzed below. In addition to depending on the plasma parameters, the experimentally obtained line shapes for spectral lines will also depend on the spectral devices and on the photographic materials. In order to determine the contribution to the broadening due to the device or, as it is frequently called, the apparatus width, it is possible to make use of a line of a plasma known to be cold. Obviously one should choose a line whose broadening as a consequence of other factors is small under the given experimental conditions.

If the experimentally determined line width of the plasma is appreciably greater than the apparatus width and if estimates show that other factors that cause broadening do not play a role, this line can be used to determine the ion temperature. In determining the ion temperature it is sufficient to measure the half-width of the line, that is to say, its width at a height equal to half the maximum intensity. The half-width is related to the temperature by the following relation:

$$\delta\lambda_D = \frac{2\sqrt{2k\ln2}}{c}\lambda_0\sqrt{\frac{T_i}{M}}$$

or

$$T_i(\text{eV}) = 1.7 \cdot 10^8 A \left(\frac{\delta\lambda_D}{\lambda_0}\right)^2, \tag{4.4}$$

where A is the atomic weight of the ion. In order to verify the validity of conclusions as to the Doppler broadening of a line it is necessary to plot the theoretical width corresponding to the measured values of $\delta\lambda$ and I_0. However, even if the experimentally obtained line shape and the line shape computed on the basis of Doppler broadening coincide it is not always possible to use Eq. (4.4) to determine the ion temperature. The Doppler broadening need not only occur as a consequence of random particle motion, but can also occur as a consequence of collective motion in separate regions of the plasma. In this case, if the spectral instrument accepts light from different portions of a plasma that is moving in a complicated way, the analysis of the line shape cannot show whether the broadening is a consequence of the thermal motion of the ions or whether it is due to collective plasma motion.

If the width of the Doppler line is comparable with the apparatus width, it is necessary to know the apparatus width in order to examine the Doppler shape [1, 2]. In general the form of the energy distribution in a line obtained at the output of a spectral device can be written in the form of an integral,

$$f(x) = \int_{-\infty}^{\infty} \xi(x - x')\,\varphi(x')\,dx', \qquad (4.5)$$

where $\varphi(x)$ is the energy distribution in the spectral line. If the line shape is determined by the Doppler effect associated with a random velocity distribution, then $\varphi(x) = Ae^{-x^2/a^2}$. The function $\xi(x - x')$ describes the contour which would be obtained at the output of the spectral instrument with monochromatic radiation incident upon it, and is called the apparatus function. The form of this function depends on the spectral device and the photographic film that is used. According to established terminology $f(x)$ is called the convolution of ξ and φ. We recall the following properties of the convolution of two functions:

$$\int_{-\infty}^{\infty} \xi(x - x')\varphi(x')\,dx' = \int_{-\infty}^{\infty} \varphi(x - x')\xi(x')\,dx';$$

$$\int_{-\infty}^{\infty}\int_{-\infty}^{\infty} \xi(x - x')\,\varphi(x')\,dx'\,dx = \int_{-\infty}^{\infty} \xi(x)\,dx \int_{-\infty}^{\infty} \varphi(x)\,dx. \qquad (4.6)$$

Unfortunately, the form of the function $\xi(x)$ can vary widely for different spectral devices and different detection methods. The nature of this function must be determined in each individual case. To approximate the apparatus function for a proper representation of the convolution we must take account of the conservation of energy for radiation within the line. Mathematically this statement is equivalent to the requirement

$$\int_{-\infty}^{\infty} \xi(x)\,dx = 1. \tag{4.7}$$

We shall now dwell briefly on typical examples of apparatus functions. If the device does not introduce additional distortion, for a narrow slit the apparatus function is determined by the diffraction on the aperture,

$$\xi_{dif} = \frac{1}{S_0}\left[\frac{\sin\frac{\pi x}{S_0}}{\frac{\pi x}{S_0}}\right]^2, \tag{4.8}$$

where $S_0 = \lambda f/D$ is the wavelength divided by the relative aperture diameter of the objective.

In the other limiting case, in which the slit is wide, if diffraction effects do not play a role, the apparatus function is $1/d$ when $|x| \leq d/2$, and vanishes when $x > d/2$. Here d is the slit width [the origin (x = 0) is taken at the center of the slit].

Very frequently the apparatus width is determined by the scattering of light in an emulsion and is of a dispersive nature,

$$\xi_{scat} = \frac{\frac{\alpha}{2\pi}}{x^2 + \left(\frac{\alpha}{2}\right)^2}. \tag{4.9}$$

In this case, if the shape of the spectral lines radiated by a cold plasma are approximated well by the dispersion formula, and if there are no other mechanisms that cause broadening of the spectral lines (Stark effect, Zeeman effect), the contribution of the Gaussian function associated with the Doppler effect can be determined by means of the Voigt function, which is the convolution of the Gaussian and the dispersion function,

TABLE 4.1. Voigt Line Shape for Various Half-Widths for Doppler and Gaussian Broadening

| Parameters | | | Ordinate in units of line height | | | | | | | | | | |
$\dfrac{\alpha_{1/2}}{\delta p}$	$\dfrac{\alpha_{\varphi/2}}{\alpha_{1/2}\sqrt{\ln 2}}$	$\dfrac{\alpha_{1/2}\sqrt{\ln 2}}{\delta p}$	0.8	0.7	0.6	0.5	0.4	0.3	0.2	0.1	0.05	0.02	0.01
			Distance from the center of the line in units of half-width δ_φ										
0.00	0.00	0.60	0.57	0.72	0.86	1.00	1.15	1.32	1.52	1.82	2.08	2.38	2.58
0.05	0.09	0.57	0.56	0.71	0.86	1.00	1.15	1.33	1.54	1.87	2.19	2.63	3.13
0.10	0.19	0.54	0.56	0.71	0.86	1.00	1.16	1.34	1.57	1.94	2.34	3.00	4.08
0.15	0.30	0.50	0.55	0.71	0.85	1.00	1.17	1.35	1.60	2.02	2.54	3.52	5.05
0.20	0.43	0.46	0.55	0.70	0.85	1.00	1.18	1.37	1.64	2.10	2.75	4.14	5.96
0.25	0.59	0.42	0.54	0.70	0.84	1.00	1.18	1.39	1.68	2.19	2.98	4.73	6.78
0.30	0.79	0.38	0.53	0.69	0.84	1.00	1.19	1.41	1.74	2.29	3.26	5.32	7.52
0.35	1.07	0.33	0.52	0.68	0.84	1.00	1.20	1.44	1.81	2.40	3.54	5.83	8.21
0.40	1.50	0.27	0.52	0.67	0.83	1.00	1.21	1.47	1.88	2.54	3.85	6.30	8.86
0.45	2.38	0.19	0.51	0.66	0.82	1.00	1.22	1.50	1.96	2.74	4.13	6.76	9.50
0.50	—	0.00	0.50	0.66	0.82	1.00	1.22	1.53	2.00	3.00	4.36	7.00	9.95

$$F(x) = \frac{\sqrt{\ln 2}}{\pi^{3/2}} \cdot \frac{\alpha_1}{\alpha_2} \int_{-\infty}^{\infty} \frac{e^{-\frac{4 \ln 2 x'^2}{\alpha_2^2}}}{(x - x')^2 + \frac{\alpha_1^2}{4}} \, dx. \tag{4.10}$$

Numerical values of the Voigt function for various parameters of the Gaussian and dispersion functions are given in Table 4.1. In order to determine the role of the Doppler effect in the broadening of spectral lines for various apparatus widths we can make use of various calculated curves [1]. In Fig. 4.1 we show the functional relation between the half-widths of the Doppler functions and the apparatus functions referred to the half-width of the convolution. It should be kept in mind that cases are possible in which the line shape deviates from a Guassian even though the only mechanism for broadening is the Doppler effect with a Maxwellian velocity distribution. The situation arises when the ion temperature changes with time. Under these conditions, at each instant of time the line shape is a Gaussian but the resulting shape, which is essentially the summation of a number of Gaussian curves, differs from a Gaussian. Correct information in this case can only be obtained through time scans of the shape of the spectral line.

In spite of the difficulty associated with the need for careful analysis of the experimental shapes of spectral lines, the Doppler effect is used widely for measuring ion temperature [3-7].

In plasma diagnostics the application of the Doppler effect is not limited to the determination of ion temperature and the energy associated with random motion of different parts of a plasma. There are a number of experimental papers (cf., for example, [8-10]) in which the Doppler displacement of the lines is used to establish the existence of rotational motion of a plasma in systems with circular symmetry. The same effect is also used to measure the rotational velocity. Thus, the final proof that the plasma in the Homopolar machine executes a rotational motion was obtained by observation of the Doppler shifts of the spectral lines in ionized carbon.

The observations is carried out along a line perpendicular to the assumed axis of rotation which passes through a region in which the rotational velocity of the plasma is a maximum. By photographing the spectrum of the plasma in the Homopolar, first

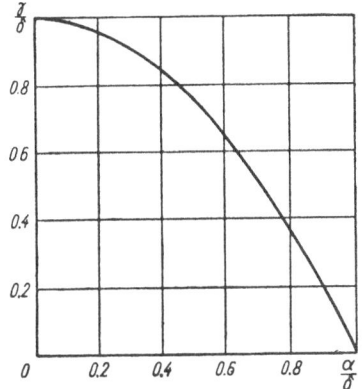

Fig. 4.1. The relation between the Doppler half-width α and the dispersion half-width γ referred to the half-width of the sweep.

in one direction of the magnetic field, and then in the opposite direction, it was possible to determine the shifts of the ion lines due to the Doppler effect. We recall that the rotation of the plasma in the Homopolar occurs in crossed **E** and **H** fields and a change in the sign of one of these leads to a change in the direction of the drift motion. As expected, the observed shifts were symmetric with respect to the lines that are observed when **H** = 0. The measured shift corresponded to a velocity of $3 \cdot 10^6$ cm/sec. Similar measurements have been carried out in other devices with rotating plasmas, in particular, the Ixion device. Here the Doppler shift was found to correspond to a velocity of $4 \cdot 10^6$ cm/sec.

By observing shifts of various spectral lines associated with neutral atoms and ions it is possible to determine the velocity of individual plasma components and, in particular, it is possible to determine whether the motion of the neutral gas is the same as the velocity of the motion associated with the ionized gas.

In order to measure the ion temperature of a plasma in a thermonuclear device it is necessary to understand the Zeeman splitting of the spectral lines in the magnetic field used to confine the plasma. If the magnetic field is so large that the Doppler width is comparable with $\lambda^2 (eH/2\pi mc^2)$, then the Zeeman effect becomes an important background effect. Although the analysis of the experimental data that is required to obtain the Doppler shape under these conditions is extremely complicated, the situation is not entirely hopeless. It is found that the problem of resolving the Doppler line shape for a plasma line in a strong magnetic field can be solved experimentally [11]. As an example of this possibility we consider the OVI 3811 Å ($2S_{1/2} - 2P_{3/2}$) line, for which the anomalous Zeeman effect is observed.

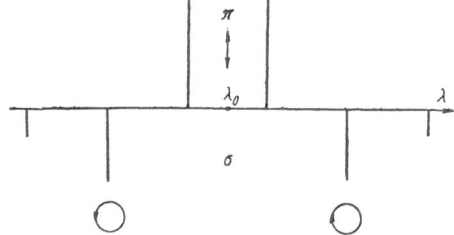

Fig. 4.2. Diagram showing the Zeeman effect
for the $2S_{1/2} - 2P_{3/2}$ transition.

The splitting of the Zeeman components in the anomalous
Zeeman effect is given by the expression

$$\Delta\lambda = (m_1 g_1 - m_2 g_2) \lambda^2 \frac{eH}{2\pi mc^2}, \qquad (4.11)$$

where m_1 and m_2 are the magnetic quantum numbers while g is
the Landé factor for Zeeman sublevels. The selection rule for
the transitions is given by $\Delta m = 0, \pm 1$. A splitting scheme in a
magnetic field for the $2S_{1/2} - 2P_{3/2}$ transition is shown in Fig. 4.2.
The line has two π components, whose emission is polarized along
the magnetic field, as well as two weak and two strong components
which are circularly polarized (σ components).

In the observation of the emission that propagates along the
magnetic field the π component is not observed since the electric
vector of this wave is directed along the field. The σ components
are observed in this case; these components are circularly polarized. The rotation of the polarization vector for the components
located on one side of the unsplit line is in the opposite direction
to the rotation of the σ components located on the other side.
The separation of the σ components with the same direction of
rotation of the polarization vector is realized by means of a
quarter-wave plate and a polarizer. After transmission through
the quarter-wave plate the circularly polarized radiation becomes
plane polarized; the planes of polarization of the components that
rotate in opposite directions are at right angles after transmission
through the quarter-wave plate. The polarizer located beyond the
quarter-wave plate only separates the σ components which are
displaced by the effective magnetic field in one direction from the

unsplit line. Since there are only two such lines for the $2S_{1/2}-2P_{3/2}$ transition, and since the intensities of these lines are very different, the contribution of the Zeeman effect to the line shape can be reduced by application of the quarter-wave plate and polarizer; a reduction of at least one order of magnitude can be obtained in this way. As a result the effect of the magnetic field on the accuracy of the measurement of the ion temperature becomes insignificant even at the high magnetic fields for which the Zeeman splitting becomes comparable with the Doppler broadening.

The inverse problem is of no less interest, that is to say, the problem of resolving the Zeeman splitting of a line broadened by the Doppler effect. In this case it is sufficient to determine the distance between lines corresponding to the right- and left-hand rotation of the plane of polarization. By knowing the distance between the split components and using the dependence of the splitting on the magnitude of the magnetic field it is possible to determine the field strength in the region from which the emission originates. One interesting feature of this method is the fact that it is free of perturbations of the plasma such as those that occur with magnetic probes. Unfortunately, up to this time this method has not received the attention it deserves in investigations of laboratory plasmas; however, it is widely used in astrophysics.

§ 4.3. Doppler Effect in Scattered Light

Determination of Electron Temperature. With the advent of lasers which generate intense beams of monochromatic radiation it is now possible to observe the Doppler effect not only in line radiation of the plasma itself, but also in the radiation from an external source which is scattered by a plasma. Under certain conditions the scattering occurs primarily on the electrons. Hence, an analysis of the lines associated with the scattered light can, in principle, provide information on the electron density as well as the electron velocity distribution. In this form the problem is extremely complicated; for this reason, we shall only consider a particular case, specifically the determination of the temperature assuming that the plasma exhibits a Maxwellian distribution.

We shall first consider the general features of scattering of a monochromatic beam characterized by a wavelength that lies in the visible region. We consider scattering on a plasma. If the wavelength of the incident light is much smaller than the mean distance between charged particles and is also smaller than the Debye radius, then the effect of the interaction of the radiation with the plasma is determined by elementary scattering events of photons on individual particles. The total cross section for this process is given by the Thomson formula

$$\sigma = \frac{8\pi}{3} \left(\frac{e^2}{mc^2} \right)^2 \tag{4.12}$$

and is $6.65 \cdot 10^{-25}$ cm^2 for electrons. We note in passing that for the red line emitted by the ruby laser ($\lambda = 6943$ Å) and for typical values of the electron temperature ($T = 20$ eV), according to Eq. (2.1) the density for which the wavelength becomes comparable with the Debye radius is $2 \cdot 10^{15}$ cm^{-3}. This value corresponds to the limiting density for which the Thomson scattering formula can still be used. At high densities the scattering cannot be considered in terms of elementary interaction events between a photon and individual charged particles in the plasma. A more precise criterion for the application of the Thomson formula will be given below. Since there exists a strong dependence of the effective cross section on the mass of the particles, the relative contribution to the scattering due to the ions is small. In our analysis we shall only consider scattering on the electrons. In work with plane-polarized light, the scattering is not isotropic and the power scattered at an angle θ with respect to the electric vector of the initial radiation is

$$W_s(\theta)d\theta = \left(\frac{e^2}{mc^2} \right)^2 W_0 \int_0^L n \; dl \cos^2\theta \; d\theta, \tag{4.13}$$

where n is the density and L is the beam path in the plasma.

Substituting the values of the constants we can show easily that for a plasma with a density of approximately 10^{15} cm^{-3} which is 10 cm long, the fraction of scattered light W_s/W_0 is approximately 10^{-8}, that is to say, the possibility of detecting this light is at the limits of the present state of the art because of the difficulty of obtaining a reasonable signal-to-noise ratio. Actually the situation is somewhat better then it appears since the half-width of the line associated with the scattered light is much greater than the half-width of the incident light; consequently, it is possible to measure the contribution from the center of the spectral line and, in this way, to reduce sharply the effect of the background scattering.

The line shape of the light scattered by the electrons is determined by the Doppler effect; however, in contrast with broadening of the lines of plasma radiation, in the present case in all formulas we use the electron velocity rather than the ion velocity. For a Maxwellian electron velocity distribution the half-width for the line associated with the scattered light is

$$\delta\lambda_s = 3.3 \cdot 10^{-3} \lambda \sqrt{T_e(\text{eV})}. \tag{4.14}$$

For the $\lambda = 7000$ Å line at a temperature of $T_e = 100$ eV the halfwidth is 230 Å, which is to say that the half-width is three orders of magnitude larger than the total Doppler half-width of the line emitted by the plasma for $T_i = 100$ eV. Thus, in addition to measuring the density by the intensity of the scattered light [12, 13], which requires an absolute measurement, it is also possible to use the scattered radiation to determine the electron temperature [14]. This large width for the detected line has certain disadvantages as well as advantages. At the extremes of the line there may be a background caused by bright emission lines of the plasma and this can make the measurement of the intensity of the scattered light very difficult. In particular, in the red region there is a bright hydrogen line Hα 6561 Å. However these difficulties are not fundamental because it is always possible to determine the temperature by using the other side of the line.

The situation is somewhat more complicated as regards the continuum radiation of the plasma; this background cannot be reduced by selection of a spectral range for detection. From the formulas for the bremsstrahlung and recombination radiation which are given below in §5.1 it follows that no existing source of radiation other than the laser provides sufficient power for studying the line shape of light scattered from a plasma. In a plasma with an electron temperature below 10 eV only present-day lasers which operate in the pulsed mode with powers of 10^7 W can provide a signal-to-noise ratio greater than unity up to values of the density for which the wavelength of the scattered light is equal to the Debye radius. The minimum density for which scattered radiation can be detected is determined by the minimum number of scattered

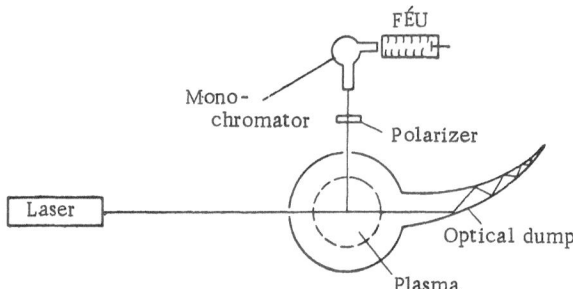

Fig. 4.3. Diagram showing the determination of electron temperature by measurement of the line shape of
scattered light.

photons that can be recorded by a photomultiplier. According to
the data in [13], the use of a double monochromator DFS-12 and a
laser with a radiated energy of 100 J provides the possibility of
working at a plasma density below 10^{12} cm^{-3}.

A typical experimental arrangement in which the electron temperature is measured by analysis of the line shape of scattered
radiation is shown in Fig. 4.3. The beam of light from the laser
is scattered in the plasma and the unscattered part (i.e., the primary part) enters an optical dump. The optical dump is a chamber
with a system of labyrinths established in such a way that the beam
can only escape from it after multiple reflection from blackened
surfaces. The detection apparatus is arranged in such a way that
the entrance slit only receives photons that have been scattered at
an angle of 90° with respect to the primary beam along the direction of the electric vector of the wave incident on the plasma (if
the light from the laser is polarized). The use of a monochromator makes it possible to isolate a narrow spectral region at some
distance from the center of the line and, in this way, to reduce
the radiation background scattered on the plasma. An additional
increase in signal-to-noise ratio cɪn be achieved by selecting
the polarized radiation scattered at a particular angle since the
light scattered by the plasma electrons is fully polarized. The
consideration has been used, for example, in [15, 16] where the
polarized light was separated from the unpolarized light by means
of a differential amplifier. The polarized portion of the scattered
radiation is due to scattering by the plasma electrons.

By recording the radiation scattered by the plasma it is possible to determine the plasma density as well as the electron temperature. However, this measurement requires a knowledge of the absolute value of the energy of the scattered radiation from a specified volume at a specified angle, in which case the value of the plasma density can be obtained directly from Eq. (4.13). In [20] a system is described in which an interference filter is used together with polaroid filters for detection of the laser radiation scattered at an angle of 90°. By using this system the authors were able to measure plasma densities of 10^{13} cm^{-3}.

At the present time this method of determining electron temperature and density from the width of the line and the intensity of the scattered light has received considerable popularity (cf., for example, [17, 19]). The measurement apparatus used in each of these published papers has its own characteristic features, but these different features are not fundamental.

The foregoing analysis is based on the assumption that we are dealing with the Doppler effect associated with the scattering of light on free electrons in the plasma. This assumption is valid only for a plasma in which the density is not very high and in which the wavelength of the laser radiation is not very large compared with the Debye radius. The condition that must be satisfied if the scattering of light is to occur by free electrons, in which case the scattering is described by the Thomson formula, is then

$$\alpha = \frac{\lambda}{4\pi \sin\frac{\theta}{2} \lambda_D} \ll 1, \tag{4.15}$$

where θ is the scattering angle and λ_D is the Debye radius. The line-shape problem for light scattered by a plasma has been considered in general form by Rosenbluth and Rostoker [21, 22]. In the limit $\alpha \ll 1$ the expression obtained by these authors becomes the formula for the line shape of light scattered by free electrons:

$$I = I_0 e^{-\frac{1}{2}\left(\frac{\Delta\lambda}{\lambda}\cdot\frac{c}{\sqrt{\frac{kT_e}{m}}}\right)^2}, \tag{4.16}$$

where c is the velocity of light.

In the other limiting case, in which the wavelength of the radiation is much greater than the Debye radius $(\alpha \gg 1)$, which is the case in a plasma of high density and low temperature, the scattering occurs not only on individual electrons, but also on fluctuations of the charge density of charges that are grouped around the ions.

Calculations that have been carried out for an isothermal plasma at densities 10^{16}-10^{18} cm^{-3} [23, 24] show that at high densities the central portion of the scattering line is a rather narrow peak which is similar in shape to the ion Doppler line. However, the wings of the line are associated primarily with the Doppler effect due to scattering by free electrons. As the density increases the fraction of the energy scattered in the central portion of the line increases rapidly.

For plasma densities of $\sim 10^{17}$ cm^{-3} the spectral power of the radiation scattered by the density fluctuations at an angle of 90° is approximately two orders of magnitude greater than the spectral density of the radiation scattered by free electrons. As the density increases, at $\alpha \sim 1$ there arises a clearly distinguished feature of the line shape of the scattered radiation. Symmetrically arranged with respect to the center of the line are two peaks which are displaced from the center by a distance approximately equal to the plasma frequency. When $\alpha \gtrsim 1$ the intensities of these peaks amount to $1/\alpha^2$ of the total energy of the scattered light. Curves of the line shapes for scattering by a dense isothermal plasma are given in [24].

In a nonisothermal dense plasma with $T_e \gg T_i$ the spectral distribution in the central portion of the line exhibits two peaks which are symmetrically arranged with respect to the frequency of the radiation of the line and the separation of the peaks from the center is determined by the ion acoustic frequency

$$\omega_{\text{acoust}} = \frac{2\pi}{\lambda} \sqrt{\frac{ZkT_e}{M}} \cdot \frac{1}{\sqrt{1 + \frac{4\pi^2 \lambda_D^2}{\lambda^2}}} \approx \frac{2\pi}{\lambda} \sqrt{\frac{ZkT_e}{M}}. \qquad (4.17)$$

Thus, in principle, this limiting case provides an additional method for determining the electron temperature of the plasma.

A general expression for the energy distribution in the spectrum of the scattered light for $\alpha \gg 1$ is rather complicated and

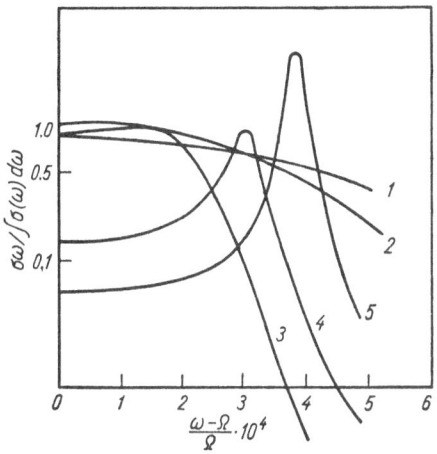

Curve	T_e, eV	T_i, eV	σ_{total}/r_0^2
1	5	50	0.812
2	5	25	0.808
3	5	5	0.500
4	25	5	0.236
5	50	5	0.360

Fig. 4.4. Line shapes for scattered light corresponding to various temperatures.

is not suitable for practical use. At the center of the line

$$\sigma(\omega) = \left(\frac{e^2}{mc^2}\right)^2 \frac{1 - \sin^2\theta\cos^2\varphi}{2\sqrt{\varepsilon}\,\Omega\sin\frac{\theta}{2}} \sqrt{\frac{Mc^2}{2\pi kT_i}} \times$$

$$\times \frac{1 + \left(\dfrac{m}{M}\cdot\dfrac{T_e^3}{T_i^3}\right)\left[1 + (K-\chi)^2\lambda_{Di}^2\right]^2}{\left[1 + (K-\chi)^2\lambda_{De} + \dfrac{\lambda_{De}^2}{\lambda_{Di}^2}\right]^2}, \qquad (4.18)$$

where ε is the dielectric constant of the plasma; Ω is the frequency of the laser radiation; φ is the angle between the projection of the wave vector of the scattered light on the plane perpendicular to the light beam and the electric vector of the transmitted wave; $\lambda_{Di} = \sqrt{\dfrac{T_i}{4\pi e^2 n}}$; $\lambda_{De} = \sqrt{\dfrac{T_e}{4\pi e^2 n}}$; K and χ are the wave vectors for the scattered wave and the probing wave. For purposes of illustration

TABLE 4.2. Scattering Cross
Section at an Angle of 30°
in Units of Solid Angle 10^{-28} cm

	Ne	Ar	Xe
Experiment	0.048	0.843	5.71
Theory	0.027	0.47	2.88

of the shape of the spectral line (4.16) for various values of T_e and T_i in Fig. 4.4 we show the results of a numerical calculation of the spectrum of the scattered radiation $\sigma(\Delta\omega/\Omega)$, where $\Delta\omega = \omega - \Omega$. These calculations apply for a plasma density of 10^{12} cm^{-3} and a wavelength of 1.7 cm and the results characterize the spectral part of the line for $\alpha \gg 1$.

At intermediate values of the density, that is to say, when $\alpha \sim 1$, it is necessary to compare the theoretical and experimental spectra for the scattered light in order to determine the density and temperature.

It is interesting to note that the total cross section, i.e., the cross section integrated over all frequencies, can always be given by the Thomson relation for order-of-magnitude accuracy. For example, in an isothermal plasma, in which the integration over frequency can be carried out by analytic methods, the total cross section for scattering for $\alpha \gg 1$ is only one-half that computed from (4.12).

A serious difficulty arises in measurements of plasma density from the intensity of the radiation scattered on electrons. This is the experimental problem of measuring the relative intensity of the incident and scattered radiation; the ratio of the quantities can be many orders of magnitude. This problem is undoubtedly easier than that of determining the absolute intensity of each of the components separately. The measurements can be simplified appreciably by using a known scattering standard which has a scattering coefficient close to that of the plasma. It is convenient to use a spectrally pure gas which does not contain dust particles as such a standard [25]. The intensity of the radiation scattered by such a gas is determined by the Rayleigh scattering.

The expression for the intensity of the light scattered from an area of 1 cm² oriented at right angles to the direction of motion of the scattered photon is

$$I(\theta) = I_0 \frac{(2\pi)^4 \alpha^2 n_0 l}{2\lambda^4 r^2}(1 + \cos^2\theta), \qquad (4.19)$$

where I_0 is the intensity of the incident unpolarized light; θ is the angle between the direction of the incident light and the scattered light; n_0 is the density of molecules in the scattering volume; r is the distance to the scattering volume; λ is the wavelength; l is the path length of the light in the gas; and α is the polarizability of the molecule, which is related to the gas density n_0 and the reference index N by the relation

$$\alpha = \frac{N-1}{2\pi n_0}. \qquad (4.20)$$

The refractive index for certain gases is given in Table 7.1 (Chapter 7). If use is made of the polarized emission of a laser for which the direction of polarization is in the same plane as the scattering direction (4.19) still applies if we omit the term unity in the bracket and multiply the right-hand side by a factor of two.

Equation (4.19) has recently received experimental confirmation [26]. The experimental results show that the value of the cross section exceeds the theoretical value by approximately a factor of two (Table 4.2). However, these data are still not to be regarded as definitive and require additional verification.

§ 4.4. Stark Broadening of Spectral Lines in a Plasma. Determination of the Ion Density

An atom located in a plasma is subjected to the electric microfields of charged particles. Hence, in addition to the Doppler broadening of the spectral lines the lines can also be broadened by the Stark effect. It should be noted, however, that the electric microfields in the plasma oscillate rapidly, changing both in magnitude and direction; thus, the determination of the line shape can be an extremely complicated problem.

In analyzing the broadening of atomic lines in a plasma it is important to remember that the characteristic frequencies of the ion electric fields are $(M/m)^{1/2}$ times smaller than the characteristic frequencies of the electron microfields. All other conditions being equal, the ion velocities and electron velocities are in the ratio $(m/M)^{1/2}$. In other words, in one time of flight of an ion the excited atom experiences $\sim(M/m)^{1/2}$ collisions with plasma electrons. Hence, the following physical picture applies: the fixed atom, located in the quasistatic electric field of the ions, experiences the effect of the moving electrons. The slowly varying electric field due to the ions leads to a Stark broadening of the atomic levels while the rapidly varying electric microfields due to the electrons cause transitions between the individual Stark levels; in addition, sharp changes in the magnitude and direction of the field cause marked changes in the phase of the emitted light.

The broadening of the individual Stark components of spectral lines caused by electrons can be described by the so-called collision theory. In this theory it is assumed that a collision changes sharply the train of emitted light waves and that the time interval between collisions can be much greater than the collision time itself. The line shape described by the collision theory is a Lorentzian,

$$I_s = \frac{\gamma}{(\omega - \omega_0 - d)^2 + \dfrac{\gamma^2}{4}}, \tag{4.21}$$

where ω_0 is the frequency and d is the displacement of the line in the electric field.

The half-width of the line is essentially equal to the reciprocal lifetime of the atom in a given excited state $\gamma = n_e \langle \sigma v_e \rangle$. The cross section for electron collisions that cause the transition from the given excited state to another state can be easily estimated from the following relation:

$$\sigma \approx \pi \varrho^2, \tag{4.22}$$

where ρ is the characteristic value of the collision parameter for which an inelastic collision is possible between the electron and the atom.

In the semiclassical approximation, the transition from one state to another under the effect of collisions is possible in the case in which the time of flight of the particle is smaller or comparable with the characteristic time for the transition between states. Transitions in an excited atom occur primarily between states with the same principal quantum number and the condition which defines the collision parameter in the hydrogen-like approximation can be written in the following form:

$$\frac{v}{\rho} = \frac{\langle d_{mm'}\rangle e}{\hbar\rho^2}.$$ (4.23)

Here $\langle d_{mm'}\rangle$ is the dipole moment of the transition for the Stark components. On the right side of this relation we have the transition frequency between sublevels with the given principal quantum number κ caused by collisions with electrons. Writing the dipole moment in the form $\kappa^2 a_0 e$ and substituting the value of the Bohr radius a_0 we obtain an expression for the collision parameter,

$$\rho \approx \frac{\kappa^2 \hbar}{Zmv}.$$

Then the cross section which determines the probability of a transition between sublevels for a given K in an excited atom can be written in the following form:

$$\sigma \approx \pi\kappa^4 \left(\frac{h}{mv}\right)^2 Z^2.$$ (4.24)

Actually, because of the long-range nature of the dipole interaction, on the right side of this relation we should add a factor $L = \ln\frac{mv_e^2}{E-E^1}$, which is approximately equal to 10.

We note now that the effects of adiabatic direction changes of the field during the transit time of the electrons make an appreciably smaller contribution to the broadening than the shift of the lines; furthermore these have no effect on the lifetime of the atom in a given excited state.

The collision theory only gives a good description of electron contribution to the broadening of the spectral line because developing it requires that the duration of a collision be much smaller than the time between collisions; thus, the criterion for the application of the collision theory can be written in the following manner:

$$\frac{n^{-1/3}}{v_e} \ll \frac{1}{n \langle \sigma v_e \rangle} \, ,$$

or, substituting the probability for a collisional transition,

$$\frac{n^{-1/3}}{v_e} \ll \frac{m^2 v_e}{n \pi \kappa^3 \hbar^2 L} \, . \tag{4.25}$$

This relation can be satisfied for ions only at extremely low densities and high temperatures in which case the Stark broadening is small and the line shape is determined primarily by the Doppler effect.

In the region of rather high densities but only modestly high temperatures the line shape due to ion effects can be described by the so-called statistical (more precise) quasistatic theory. If the lifetime of the excited atom is shorter than the characteristic time for a change in the ion microfield an excited atom in a plasma acts as though it were "photographing" a fixed pattern of the distribution of ion electric microfields; under the effect of the electron collisions the atom makes a transition to another excited state before this pattern can change. It is easy to show that the criterion for the validity of this approximation is the inverse of that given in (4.25), where we now use the ion velocity on the left side of the inequality,

$$\frac{n^{1/3}}{v_i} \gg \frac{m^2 v_e}{n \pi \kappa^4 \hbar^2 L} \, . \tag{4.26}$$

This criterion is satisfied better at higher plasma densities and lower temperatures.

The application of the quasistatic theory requires a rather large width for the individual Stark level ($\gamma = n \langle \sigma v_e \rangle$), hence the quasistatic approximation only describes portions of the line that lie far from the center $\omega - \omega_0 > \gamma$. In this case the central portion of the line $\omega - \omega_0 \le \gamma$ must be described by collisional broadening due to electrons.

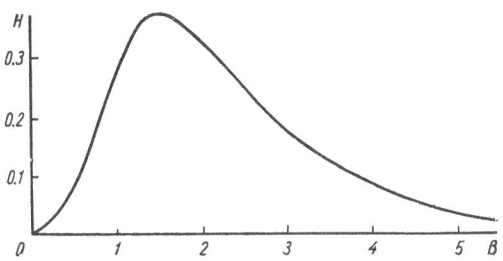

Fig. 4.5. Curve of the Holtsmark function.

Criteria for the validity of the quasistatic broadening were formulated relatively recently by G. V. Sholin and L. P. Kudrin [27] although the calculation of the probabilities for the distribution of ion microfields in the plasma was carried out in 1908 by Holtsmark (cf., for example, [28, 29, 30]). It was shown by Holtsmark that the probability distribution of microfields can be written in the form

$$H(\beta) = \frac{2}{\pi\beta} \int_0^\infty v \sin v e^{-\left(\frac{v}{\beta}\right)^{3/2}} dv, \tag{4.27}$$

where

$$\beta = \frac{E}{E_0}; \tag{4.28}$$

$$E_0 = 2.61 \, en^{2/3}.$$

The physical significance of E_0 is simple: the mean distance between plasma particles is $n^{-1/3}$ while the field of a point charge is inversely proportional to the square of the distance. The Holtsmark function is shown in Fig. 4.5. In the limiting cases of high and low values of the argument the function $H(\beta)$ can be approximated by series, the first of which describes the remote wings of the line

$$H(\beta) = 1.496 \, \beta^{-5/2} (1 + 5.107 \, \beta^{-3/2} + 14.43 \, \beta^{-3} + \ldots), \quad \beta \gg 1, \tag{4.29}$$

$$H(\beta) = \frac{4}{3\pi} \beta^2 (1 - 0.463 \, \beta^2 + 0.1227 \, \beta^4 + \ldots), \quad \beta \ll 1. \tag{4.30}$$

TABLE 4.3. Stark Constants for Hydrogen Lines

Line	H_α	H_β	H_γ	H_δ
$\lambda \overset{\circ}{A}$	6 563	4861	4 102	3 970
$\overline{\alpha}$	3.9	10.5	21	25
$\overline{A}$	$1.4 \cdot 10^{-10}$	$2 \cdot 10^{-10}$	$2.9 \cdot 10^{-10}$	$3.1 \cdot 10^{-10}$

The line shape in the region of applicability of the statistical analysis is determined on a frequency scale in the following way:

$$I_{st}(\omega) = \frac{1}{\Delta\omega_0} H\left(\frac{\omega - \omega_0}{\Delta\omega}\right). \qquad (4.31)$$

In this formula, for the linear Stark effect we have

$$\Delta\omega_0 = 2.61 \, \alpha n^{2/3}. \qquad (4.32)$$

In contrast with the Doppler shape the Holtsmark line shape falls off slowly in the remote wings, following the relation $\beta^{-5/2}$. For lines whose broadening is due simultaneously to statistical fields and the Doppler effect, extrapolation of the far wings of the line by the power relation (4.27) makes it possible to determine $\Delta\omega_0$ and, consequently, the plasma density. At large distances from the center of the line the Doppler contribution becomes negligibly small because of the exponential decay. In those cases in which it is not fruitful to try for high accuracy, it is sufficient to limit the analysis to the first term in the expansion in the Holtsmark function.

The half-width of a line which is broadened by the statistical field of ions can be obtained from the formula

$$\delta\lambda_H = \overline{A} n^{2/3} (\overset{\circ}{A}). \qquad (4.33)$$

Values of the constants A and α averaged over all components for hydrogen lines in the visible region of the spectrum are given in Table 4.3.

Although the thermal motion of the ions is of interest from the point of view of plasma physics, in most problems it is not fundamental. In those cases in which the criteria in (4.26) are violated, but the deviation from the quasistatic theory is still not large, we can take account of the thermal motion of the ions in the form of a small correction to the Holtsmark formula. The complete analysis of the line shape for hydrogen lines that are broadened as a consequence of the interaction of the radiating atom with a large number of ions executing random motion has been given in the adiabatic approximation by V. I. Kogan [31],

$$K(\omega) = \frac{1}{\Delta\omega_0}\left\{H\left(\frac{\omega-\omega_0}{\Delta\omega_0}\right) + \pi^2 h^{2/3} S\left(\frac{\omega-\omega_0}{\Delta\omega_0}\right)\right\}. \qquad (4.34)$$

Here, $h = n\left(\frac{a}{v}\right)^3 \gg 1$ is a dimensionless parameter; $v = \sqrt{\frac{2kT}{M}}$.

The second term, which represents the correction due to the thermal motion, vanishes when $h \to \infty$. The formula is valid only when the correction term is small compared with the principal term, as is always the case in the wings of the line.

In the limiting cases the function $S(\beta)$ is given by

$$S(\beta) \approx \frac{5}{256\pi}\sqrt{\frac{2}{\pi}}\left(\frac{15}{4}\right)^{5/3}\beta^{-7/2}, \quad \beta \gg 1;$$

$$S(\beta) \approx \frac{2}{135\pi^2}\left(\frac{15}{4}\right)^{5/3}\Gamma\left(\frac{1}{3}\right)\left[-1 + \frac{25}{7}\cdot\frac{2}{\sqrt{2\pi}}^{1/6}\Gamma\left(\frac{5}{6}\right)\beta^2\right], \beta \ll 1.$$

$$(4.35)$$

At large values of h the function S falls off more rapidly than H with increasing $\omega - \omega_0$; thus, the Holtsmark formula is best applied far from the center of the line, that is to say, in the case in which the plasma density can be determined by extrapolation of the line shape to large distances from the center of the line through the power-law relation $H(\beta) = 1.5\beta^{5/2}$.

The line shape that arises under the effect of simultaneous collisional broadening by electrons and the quasistatic field of the ions can be obtained by integration of the expression

$$I(\omega) = \int I_i(\omega') I_s(\omega - \omega')\,d\omega'. \qquad (4.36)$$

More general expressions for the line shape have been given in papers by Griem, Kolb, and Shen [32-34], who have also carried out rather comprehensive numerical calculations which make it

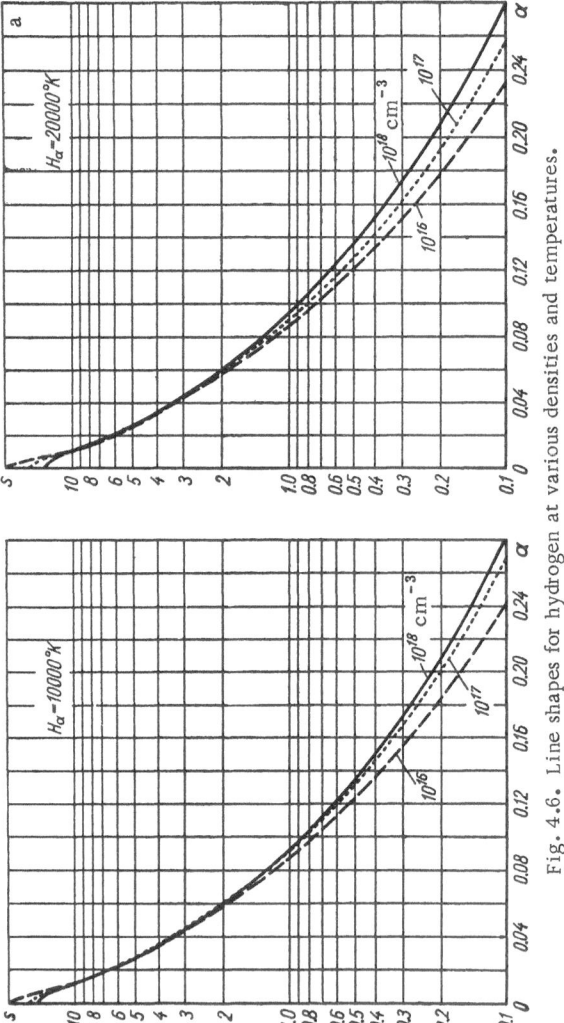

Fig. 4.6. Line shapes for hydrogen at various densities and temperatures.

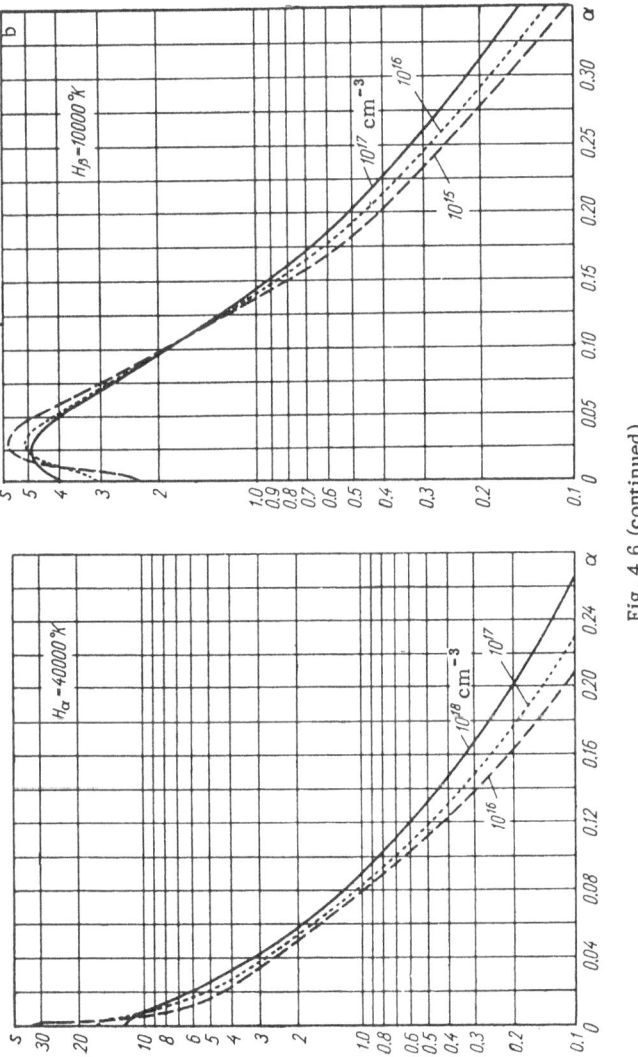

Fig. 4.6 (continued).

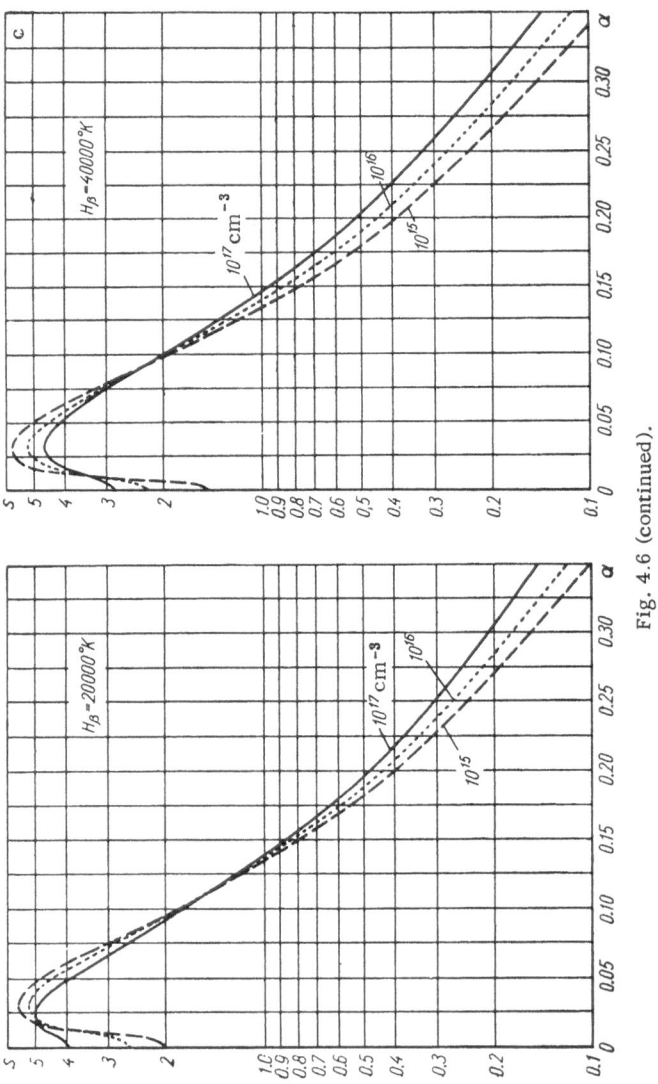

Fig. 4.6 (continued).

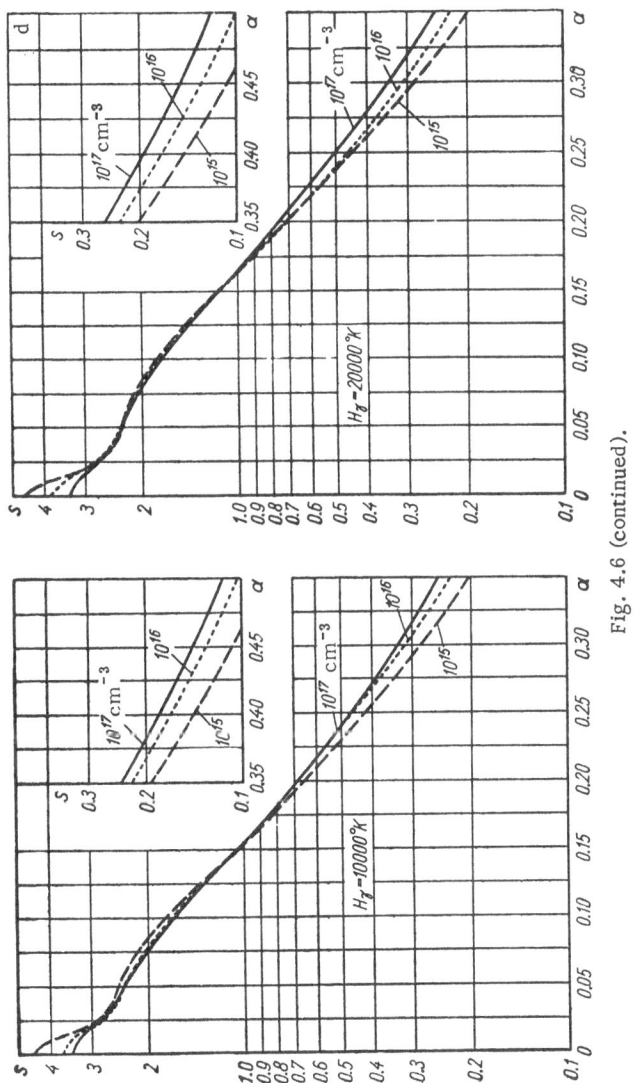

Fig. 4.6 (continued).

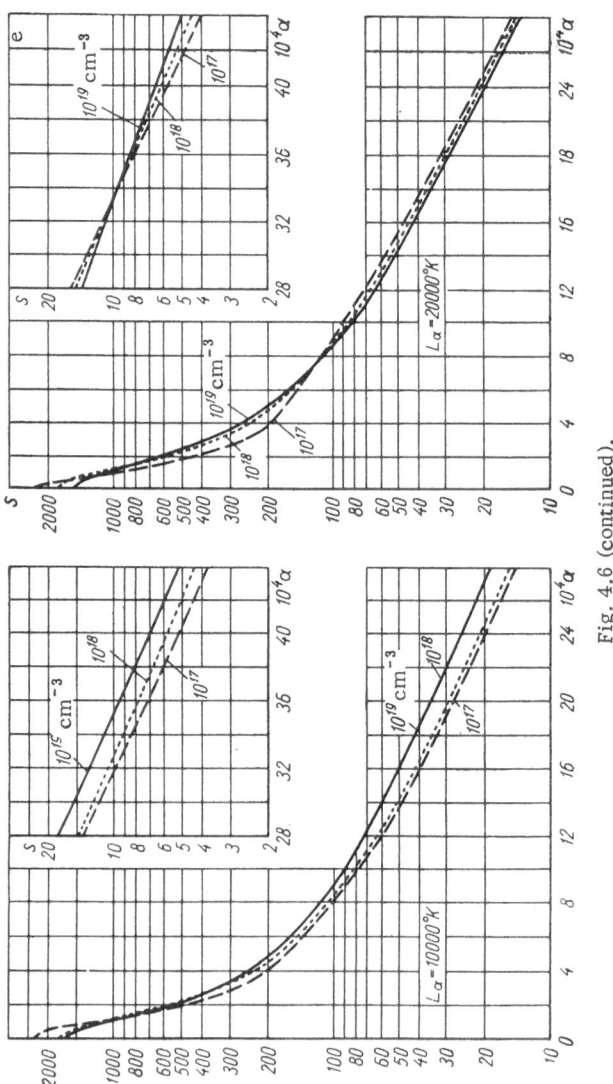

Fig. 4.6 (continued).

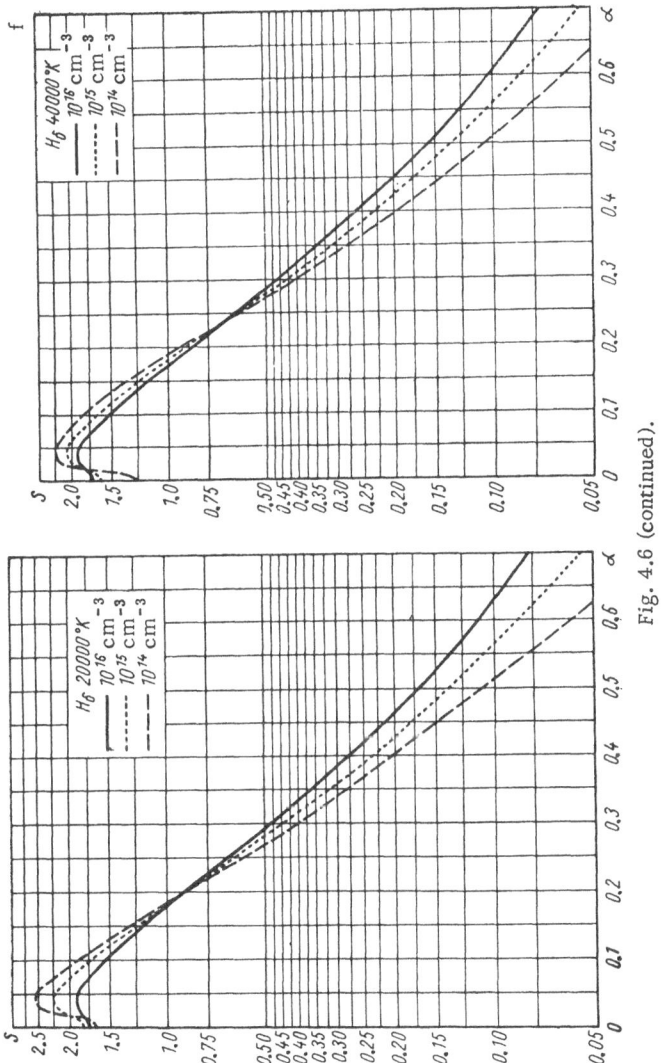

Fig. 4.6 (continued).

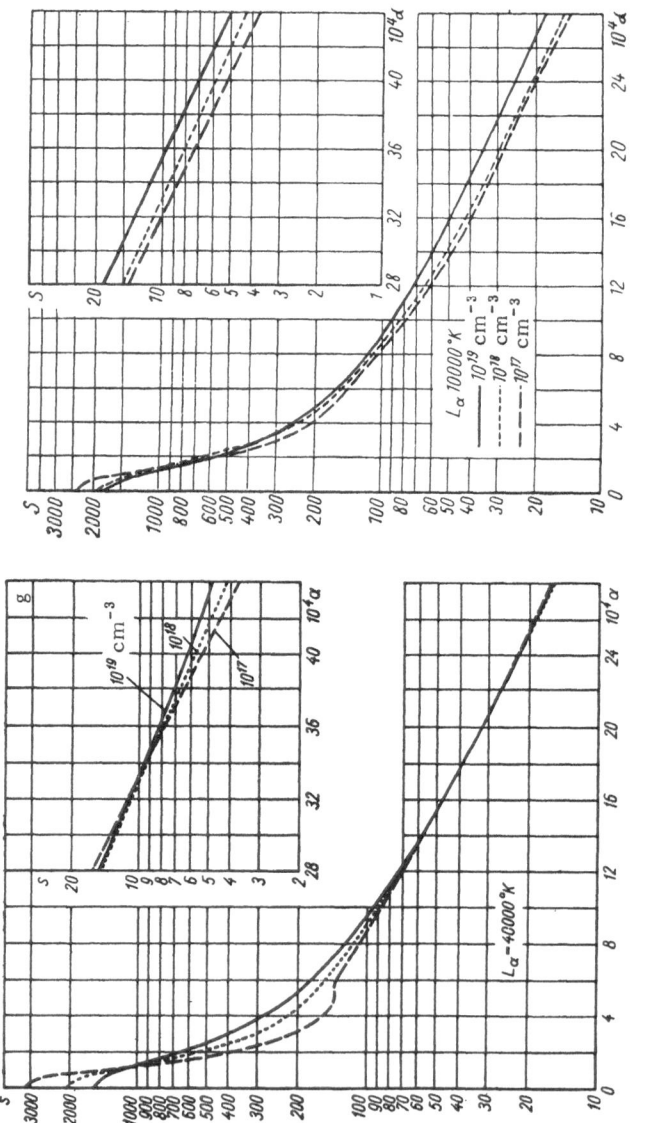

Fig. 4.6 (continued).

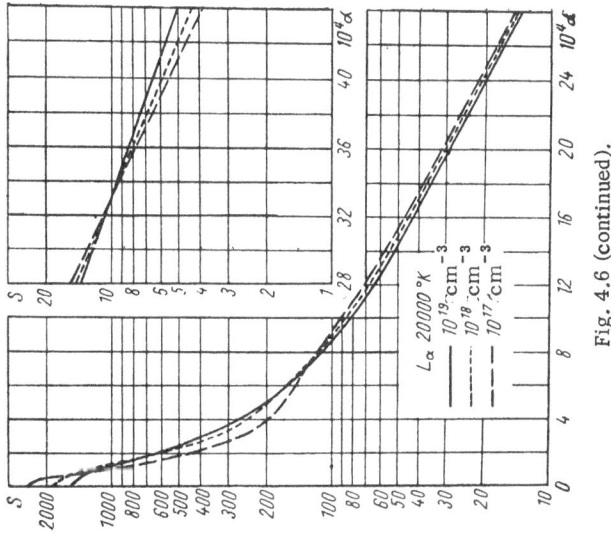

Fig. 4.6 (continued).

possible to use the theoretical results for plasma diagnostics. The calculations are carried out for various densities in the range 10^{14}-10^{18} cm^{-3}, and for temperature up to 80,000°K.

The results of the calculations for the Balmer lines and the Lyman lines in hydrogen and certain lines in helium are shown graphically in Fig. 4.6. Along the horizontal axis is plotted the quantity $\alpha' = \frac{\lambda - \lambda_0}{2.61 \; en^{2/3}}$. The density is taken in cm^{-3} and the wavelength in angstroms. Along the ordinate axis is plotted the function $s(\alpha') = \frac{2\pi c}{\lambda^2} E_0 I \left(\frac{2\pi c E_0 \alpha'}{\lambda^2} \right)$, which satisfies the normalization condition $\int s(\alpha') d\alpha' = 1$. Within the limits of the section of the line of interest here, $\lambda - \lambda_0 \ll \lambda$, it can be assumed that the function S for a given line depends only on the distance from the line center. All of the calculations are carried out neglecting the Doppler broadening since the effect of this mechanism can only be important at the center of the line. If necessary the convolution of the line shape shown on the figures and the Doppler shape can be obtained graphically.

The calculations of Griem, Kolb, and Shen make it possible to obtain a more accurate analysis of the wing of the line than is the case for the Holtsmark function. According to the calculations of these authors the wings can be extrapolated through the use of the expression

$$I(\alpha') = H(\alpha') \{ 1 + R(n, T) \sqrt{\alpha' E_0} \}, \qquad (4.37)$$

where H is the Holtsmark function, which is proportional to $(\lambda - \lambda_0)^{-5/2} \cdot R(nT)$ in the wings. The quantity $R(nT)$ is a correction computed in [33] for the first term in the Lyman and Balmer series in the density range 10^{10}-10^{18} cm^{-3} and the temperature range 0.5-4.0 eV. Over the entire range of initial conditions the quantity R is of the order of unity. Thus, the intensity of the lines in the remote wings should vary as $(\lambda - \lambda_0)^{-2}$.

Experiments carried out by V. F. Kitaev and N. N. Sobolev [35] show excellent agreement between theory and experiment.

In concluding our analysis of the Stark broadening of spectral lines we shall dwell briefly on one other method of determining plasma density which has been used successfully in astrophysical research. The idea of the method lies in the fact that the broadening of the lines leads to a degeneracy of the line spectrum into a continuum spectrum as a result of the superposition of lines one upon the other close to the limit of the series. This change can be observed clearly in lines of the hydrogen series. Actually, the distance between neighboring unperturbed levels of the hydrogen atom falls off as $1/\kappa^2$ with increasing principal quantum number while the Stark width increases as $\kappa^2 n^{2/3}$. Consequently, a given value of the plasma density n corresponds to a limiting quantum number κ for which the levels can no longer be regarded as discrete [35].

The relation between the plasma density and the principal quantum number of the upper level of the last line of the Balmer series, beyond which the Balmer series becomes a continuous spectrum [36], is usually called the Inglis–Teller formula,

$$\ln n = 23.46 - 7.5\log\kappa. \qquad (4.38)$$

Here $n = n_i + n_e$.

The basic difficulty in the measurement of plasma density through the Inglis–Teller method lies in the need for detecting extremely weak lines of transitions from levels with high quantum numbers. In a typical laboratory plasma this method becomes inconvenient at plasma densities $n \lesssim 10^{14}$ cm^{-3}. Since the higher levels of any given element are essentially hydrogen-like, the conversion of line radiation into a continuous spectrum is also observed close to the limits of the series of other elements. Experiments carried out in helium and magnesium [37, 38] show that Eq. (4.33) can be used to determine the density with an accuracy approximating a factor of two.

In addition to being observed in hydrogen atoms, the linear Stark effect is also observed in hydrogen-like ions. The constant Stark effect for a highly ionized hydrogen-like ion is much smaller than in hydrogen itself; hence these ions are not used very widely for the determination of density through the broadening of spectral lines. The numerical value of the constant α for a level with principal quantum number κ can be estimated from the formula

$$\alpha = \frac{\kappa(\kappa-1)}{Z}.$$

The Stark components of spectral lines of nonhydrogen-like atoms in a constant electric field exhibit a displacement which is

proportional to the square of the electric field, the so-called quadratic Stark effect. However, the displacement is much smaller
than that of components of hydrogen-like atoms. Thus, for example, from the available tabulated data [2] it follows that even in an
electric field of 10^5 V/cm the Stark effect for lines of nonhydrogen-
like atoms in the visible portion of the spectrum is much weaker
than for hydrogen lines. We recall that an average atomic field is
approximately 10^5 V/cm, which corresponds to a plasma density
higher than 10^{17} cm^{-3}.

In the field of the ion, the quadratic Stark effect shifts the frequency of the emitting atom by

$$\Delta\omega = \frac{\alpha_4}{R^4}, \tag{4.39}$$

where α_4 is the constant that appears in the quadratic Stark effect
while R is the distance to the ion. The width of a line in the region of applicability of the statistical broadening by ions is

$$\delta\omega = \alpha_4 (2.6)^2 n^{4/3}. \tag{4.40}$$

The criterion for the applicability of statistical broadening for the
quadratic Stark effect is of the form

$$h \equiv n \left(\frac{\pi}{2} \cdot \frac{\alpha_4}{v} \right) \gg 1. \tag{4.41}$$

Since the value of the constant α_4 lies within the limits 10^{-12}-
10^{-15}, for most cases of interest we find $h \ll 1$ so that the width
of the line is given by the collisional theory

$$\delta\omega_s = 11\alpha_4^{2/3} v^{1/3} n. \tag{4.42}$$

In an isoelectronic sequence of multiply charged ions the con-
stant α_4 falls off rapidly with increasing ion charge ($\alpha_4 \sim Z^{-3}$); for
this reason, the use of the quadratic Stark effect for the deter-
mination of density in a high-temperature plasma is not of great
practical interest. On the other hand, the expressions that have
been given here, (4.36) and (4.38), can be useful in cases in which
it is necessary to estimate the contribution to line broadening due
to different effects which distort the Doppler shape. The absolute
values of the shifts and broadening have not been investigated to
any great extent either in ionized or neutral atoms. The only exception in this regard applies to several lines of neutral oxygen

which lie in the visible region of the spectrum [39], and certain
lines in helium [40], cesium [41], and argon [42]. However, the
rapid burn up of a neutral gas makes it impossible to use them
(as is the case for hydrogen lines) in the diagnostics of high-temperature plasmas.

In considering the influence of the Stark effect on the width of
spectral lines in a plasma we have proceeded from the assumption
that the broadening occurs only by virtue of the electric fields of
charged particles. This assumption is valid in a quiescent plasma,
in which the turbulent electric fields are small compared with the
particle microfields. In the opposite limiting case of a highly tur-
bulant plasma $\left(\frac{\widetilde{E}^2}{8\pi} \sim nkT\right)$ the electric fields associated with high-
frequency oscillations can play a predominant role. Hence the
Stark broadening of lines in a turbulent plasma in which $\widetilde{E} > 2.6 \ en^{2/3}$
is more a measure of the turbulence than of the density.

§ 4.5. Time Scan of a Spectral Line

In controlled thermonuclear research it is almost always important to obtain data concerning the changes of plasma parameters
with time. A knowledge of the time behavior of the temperature
and density is required for an understanding of various processes.
Moreover, these data are also important for checking measurements since an experimental curve which coincides with the assumed functional dependence and does not contradict other experiments can always be taken to support the validity of the results. From the discussion given in the sections above it follows
that an investigation of the shape of a spectral line yields the possibility, under actual experimental conditions, of determining various plasma parameters. Hence, the development of methods which
make it possible to trace the time variation in the shape of a spectral line can be extremely important in plasma diagnostics.

It is easiest to examine the features of the change in the line
shape using a monochromator with a wavelength resolution which
is appreciably smaller than the width of the spectral line being
examined. By repeating the measurement process many times
(here we are assuming pulsed discharges) and isolating successive
narrow regions within the limits of the line it is possible to obtain
fairly reasonable experimental data with which one can determine
the line shape at any given instant of time [43]. Obviously this im-

plies that the detection apparatus at the output of the monochromator (photomultiplier and amplifier) has the required resolution time.

Unfortunately, the sequential detection of different portions of
a line can be used only if the processes in the plasma show good
reproducability so that the time variation of the line shape and
intensity is repeated in each discharge. Unfortunately, this kind
of plasma behavior is not observed very frequently so that the use
of a monochromator which records sequentially different narrow
portions of the line is not applicable in the majority of problems.
Another serious shortcoming of this method is its complexity and
the difficulty of the measurements. The change in the line width
in time can be estimated roughly by means of two photomultipliers,
one of which records the line intensity at the center and the other
the intensity in the wing. However, the information obtained in
this way is not very precise so that it is frequently not possible to
obtain a unique interpretation of the results.

The most popular method for obtaining the time behavior of
the line shape is optical scanning by means of a rotating mirror
(cf., for example, [44]). This method does not require good repro-
ducibility but it does not have the high sensitivity characteristic
of methods that use photoelectric detection; for this reason it can
only be used in work with bright lines.

A narrow, well-collimated beam of light emitted by a plasma,
is reflected from a mirror and impinges on a slit of a spectrograph. The mirror rotates with a high velocity about an axis which
is oriented in such a way that the luminous spot is moved along the
slit of the spectrograph, successfully illuminating different portions. The illuminated portion of the slit is projected through the
prism of a spectral device onto a photographic film and provides
the image of the spectrum corresponding to a given position of the
rotating mirror; consequently this process reproduces a spectrum
corresponding to a given instant of time. In order to start the exposure of the film at the instant of time corresponding to the desired phase of the discharge it is necessary to trigger the device
at a time corresponding to the desired angle of rotation of the
rotating mirror. The synchronization can be realized by means
of magnetic pickup consisting of a small coil; a small voltage is
excited in the coil when it is penetrated by a magnet that rotates

with the mirror. If a high-speed camera SFR is available the spectrum of a discharge swept in time can be obtained using the mirror in this device together with the control system and synchronization unit of the SFR. The rotational velocity of the mirror in an SFR is 75,000 rpm, which makes it possible to obtain a time resolution better than 1 μsec.

One of the most convenient devices that provides the possibility of obtaining the line shapes of a number of lines in a spectrum at different times is the electron-optical image converter. Unfortunately the high cost of such apparatus represents a serious obstacle in the application of the image converter in plasma research. These are only a small number of papers in the literature in which the authors have had the opportunity of using image converters for obtaining the time development of spectral lines [45-47].

The image of a line in the focal plane of the spectral device is projected on the photocathode of the imager converter EIC. The image formed on the photocathode provides centers of emission of photoelectrons which are produced under the effect of the light incident on the photocathode. These are projected, by means of an electron-optical system, onto a fluorescent screen which produces an intensified optical image of the portion of the spectrum being investigated. In cases in which objects with low brightness are investigated it is possible to use an image converter which has several stages of gain. The deflection system in the image converter is located between the photocathode and the first screen. Because of the low inertia of the electron beam, in principle it is possible to obtain a time resolution of the order of 10^{-14} sec. In devices being used at the present time it has been found possible to achieve a resolution time of approximately 10^{-12} sec, which is several orders of magnitude higher than the resolution that can be obtained with a rotating mirror, this figure being approximately 10^{-9} sec.

A noteworthy feature of the image converter is the possibility of synchronous detection of optical and electrical characteristics of the discharge. This feature can be provided by simultaneous provision of a triggering pulse to the trigger input of an oscilloscope and the deflection plates of the image converter. In this way it is possible to provide a synchronization accuracy of the order of 10^{-10} sec.

Image converters have additional advantages as compared with other methods of obtaining swept spectra. These are the extremely high signal-to-noise ratio as compared with photomultipliers, and the higher (approximately an order of magnitude) quantum efficiency as compared with photographic materials. In the detection of weak light signals the advantage of the image converter is enhanced by the fact that the photomultiplier detects with equal probability all electrons emitted from the photocathode, whether they are characteristic of the dark background or appear as a result of photons that strike the photocathode from the region being investigated. In the image converter, on the other hand, the background is not due to all electrons arising from spurious causes, but only those which strike a given resolution element of the image of the emitting object. The high sensitivity of the image converter, as compared with that of photographic film, does not give a complete picture of the usefulness of the EIC since a comparison of the sensitivities alone can frequently lead to erroneous conclusions. If one speaks about obtaining two identical minimum contrast images the total advantage of the EIC is determined by the ratio of the quantum yields of the photocathode in the image converter to that of the photoemulsion; according to the data of various authors this ratio varies from 10 to 100.

However, in work with an object which emitts weak and short light pulses, that is to say, under conditions in which a photographic film is not at all usable, because of its low background level and wide dynamic range the EIC makes it possible to detect a light flux which is three or four orders of magnitude weaker than that which would be detected through the use of a photographic emulsion. An image of an object can be obtained at essentially the quantum level of brightness. Obviously, this image cannot exactly reproduce all the features of the object; however, there is no choice since the limitation of information on the object lies in the low level of its emission. In the scanning of weak spectral lines the image obtained against the quantum background obviously does not permit a detailed analysis of the line shape. However, it does yield the possibility of judging variation in linewidth when other methods cannot be used at all.

Some of the basic physical principles that are used in the design of image converters have been given in a review [48] and also in collection of papers by foreign authors [49].

It is possible to obtain the line shape directly on the screen of an oscilloscope through the use of a special two-stage converter [50] which is called a double converter. The image of the line is focused on the photocathode of the first stage. Then, by means of an electrostatic lens, the electron image is transferred to the second stage of the converter, amplified, and transferred to a screen with a slit, beyond which there is an electron amplifier. The second stage has deflection plates which displace the electron image of the spectral line across the slit. A fraction of the electrons, proportional to the intensity of the spectral interval, it is defined by the width of the slit and recorded by the multiplier. The writing time for scanning a line in a double converter can be less than a microsecond.

If an EIC is not available for sweep and if the spectral lines are not too weak, satisfactory results can be obtained through the use of multilayer light pipes [51, 52]. In essence these represent the result of further development of the method of sequential measurement of individual portions of a spectral line by a monochromator together with photoelectric detection. The use of the multilayer light pipe is not restricted to the visible; it can also be used in the ultraviolet region of the spectrum in conjunction with suitable fluorescent materials.

The multilayer light pipe is fabricated from individual glass filaments which are arranged in straight lines parallel to the input slit of the spectral device. The transmission of the radiation through a light pipe from the spectrograph to the photomultiplier is based on total internal reflection at a boundary between two media which have different refractive indices. In the fabrication of multilayer light pipes use is made of a glass filament surrounded by a shell of glass with refractive index smaller than the refractive index of the core. The core of this structure then plays the role of a light pipe; the external layer of the light pipe, which has a smaller optical density, provides total internal reflection at a given solid angle and thus prevents the escape of radiation outward. Thus, the radiation is transported from one filament to another, where different filaments are joined together.

The starting point of a bundle of filaments intended for transport of radiation in a given narrow spectral range are accurately joined together in a unit and fastened with an epoxy cement. Sev-

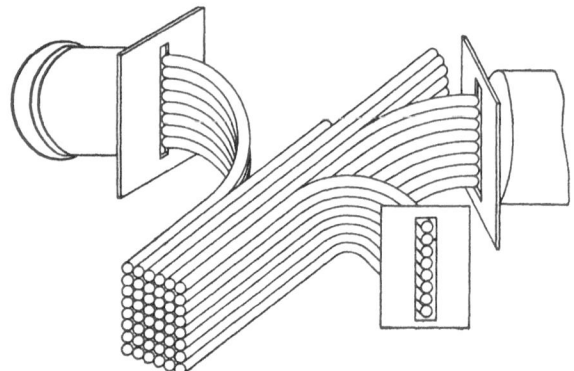

Fig. 4.7. Application of light pipes for measuring spectral
line shapes.

veral such units can be joined together and then used to form a
single assembly. After cementing, the face of the assembly of filaments is faced off and forms the detection surface on which the
image of the spectral lines is projected. In measurements of the
lines in the ultraviolet region a luminescent layer is applied to
the surface that has been faced off. The filaments in each unit
are oriented parallel to the spectral lines and are grouped in individual bundles. The opposite ends of these bundles are in optical contact with the photocathode of one of the multipliers. Thus
each of the photomultipliers records the flux of luminous energy
within a given spectral range assigned to that bundle, the width of
which is equal to the diameter of the glass filament. A schematic
diagram of a system developed for resolution of a spectral line in
time by means of fiber optics is shown in Fig. 4.7. Using 10-12
bundles of light pipes and the same number of photomultipliers it
is possible, with good accuracy, to construct the line shape for
each instant of time.

The measurement of lines characterized by a symmetric line
shape can be simplified. For example, when the basic contribution
is Doppler broadening for randomly moving atoms it is sufficient
to use one half the number of photomultipliers by investigating only
one side of the spectral line. A device that resolves a spectral line
by means of fiber optics is rather complicated since it includes
several photomultipliers, stable amplifiers, and oscilloscopes.
Nonetheless, if an EIC is not available, spectral scanning by means

of fiber optics is undoubtedly one of the most reliable and highly sensitive methods. A somewhat different approach is the idea of sweeping the spectral line by a method described in [52], in which the light pipes are wedges made of Plexiglas. In recent years attempts have been made to use fiber optics for sweeping line shapes at the output of high-resolution devices such as the Fabry – Perot interferometer.

In concluding our review of the most popular methods of obtaining time-scanned line shapes we must mention one other method: this method makes use of a Fabry-Perot interferometer in which the distance between mirrors is varied rapidly [53-55]. For a fixed position of the mirrors tne beam of light for a given spectral line forms a system of interference rings, each of which corresponds to one interference order.

For photoelectric detection of the intensity in a narrow spectral range it is necessary to focus a portion of the ring on the input slit of a photoelectric device. It will be evident that the height of the slit must be appreciably smaller than the radius of the ring. If the width of the slit is small enough, it is possible to choose only a narrow spectral range. If we now assume that the distance between mirrors varies in time, the radius of the ring will also vary. The radius is related to the distance between mirrors by the expression

$$R = \frac{\lambda f^2}{2L \, \Delta R},$$ (4.43)

where f is the focal distance of the system; ΔR is the distance between rings; L is the distance between the plates.*

Thus, if the distance between the mirrors is changed the photoelectric detector will successively record the intensity in different spectral regions. In other words, for a fixed shape and height of the spectral lines, in the oscilloscope recording the voltage at the output of a photomultiplier will show the shape of the given spectral line. This method is convenient for determining the line shape of plasma radiation in a plasma whose parameters

---

*Methods of analyzing data obtained with a Fabry-Perot interferometer and the theory of operation of the device are given in the book by Tolansky [56].

vary with time. However, the scanning time must be much smaller than the characteristic time for a change in the shape of the line being studied.

In the investigation of high-speed processes, for several years use has been made of a Fabry-Perot interferometer that contains a barium titanate element to hold the mirrors. By applying voltage pulses to the cylindrical surfaces of a ring covered with a layer of metal, which is essentially the plate of a condenser (in which the dielectric is the barium titanate), it is possible to change the mirror separation of the interferometer by virtue of the piezoelectric effect. The minimum scan time for this device is determined by the characteristic mechanical frequency of the barium titanate ring, and is of the order of 10^{-5} sec.

References

1. S. G. Rautian, Usp. Fiz. Nauk, 66:475 (1958) [Sov. Phys. — Usp., 1(2):245 (1958)].
2. C. W. Allen, Astrophysical Quantities, Oxford Univ. Press (1960).
3. S. Yu. Luk'yanov, et al., in: Proc. 2nd International Conference on the Peaceful Uses of Atomic Energy, Geneva, 1958.
4. A. N. Zaidel', et al., Zh. Tekh. Fiz., 30:1422 (1960) [Sov. Phys. — Tech. Phys., 5(12):1346 (1961)].
5. A. N. Zaidel', G. M. Malyshev, and E. A. Shreider, Zh. Tekh. Fiz., 31:129 (1961) [Sov. Phys. — Tech. Phys., 6(2):93 (1961)].
6. G. N. Harding, et al., in: Proc. 2nd International Conference on the Peaceful Uses of Atomic Energy, Geneva, 1958.
7. V. D. Pis'mennyi, I. M. Podgornyi, and Sh. Suckewer, Zh. Eksp. Teor. Fiz., 43:2008 (1962) [Sov. Phys. — JETP, 9(6):1416 (1963)].
8. O. A. Anderson, et al., in: Proc. 2nd International Conference on the Peaceful Uses of Atomic Energy, Geneva, 1958.
9. K. Boier et al., in: Proc. 2nd International Conference on the Peaceful Uses of Atomic Energy, Geneva, 1958.
10. A. A. Besshaposhnikov, et al., Zh. Tekh. Fiz., 36:1211 (1966) [Sov. Phys. — Tech. Phys., 11(7):898 (1967)].
11. A. Eberhagen, et al., Z. Naturforsch., 202:1375 (1965).
12. T. P. Hughes, Nature, 194:268 (1962).
13. A. N. Zaidel', G. M. Malyshev, and G. V. Ostrovskaya, in: Plasma Diagnostics [in Russian], Gosatomizdat, Moscow, 1963, p. 31.
14. V. V. Korobkin, in: Plasma Diagnostics [in Russian], Gosatomizdat, Moscow, 1963, p. 36.
15. E. Funfer, B. Kronast, and H. J. Kunze, Phys. Lett., 5:125 (1963).
16. G. Fiocco and E. Thomson, Phys. Rev. Lett., 10:89 (1963).
17. G. M. Malyshev, Zh. Tekh. Fiz., 35:2129 (1965) [Sov. Phys. — Tech. Phys., 10(12):1633 (1966)].

18. W. E. R. Davies and S. A. Ramsden, Phys. Lett., 8:179 (1964).
19. H. J. Kunze, et al., Phys. Rev. Lett., 11:42 (1964).
20. S. E. Swarz, Appl. Phys., 36:1836 (1965).
21. M. N. Rosenbluth and N. Rostoker, Phys. Fluids, 5:776 (1962).
22. V. D. Shafranov, in: Reviews of Plasma Physics, Consultants Bureau, New York, 1967, Vol. 3.
23. E. E. Salpeter, Phys. Rev., 120:1528 (1960).
24. E. T. Gerry and R. M. Patrick, Phys. Fluids, 8:210 (1965).
25. G. M. Malyshev, et al., Dokl. Akad. Nauk SSSR, 168:554 (1966) [Sov. Phys. − Dokl., 11(5):441 (1966)].
26. T. V. George, et al., Phys. Rev., 137A:369 (1965).
27. L. P. Kudrin and G. V. Sholin, Dokl. Akad. Nauk SSSR, 147:342 (1962) [Sov. Phys. − Dokl., 7(11):1015 (1963)].
28. I. I. Sobel'man, Introduction to the Theory of Atomic Spectra [in Russian], Fizmatgiz, Moscow, 1963.
29. H. Margenau and M. Lewis, Rev. Mod. Phys., 31:56 (1959).
30. V. Weiskopf, Usp. Fiz. Nauk, 13:596 (1933).
31. V. I. Kogan, Plasma Physics and the Problem of a Controlled Thermonuclear Reaction, Pergamon, New York, 1958, Vol. 4.
32. H. R. Griem, et al., Phys. Rev., 125:177 (1962).
33. H. R. Griem, Astrophys. J., 132:883 (1960).
34. H. R. Griem, A. C. Kolb, and K. U. Shen, Phys. Rev., 116:4 (1959).
35. V. F. Kitaeva and N. N. Sobolev, Dokl. Akad. Nauk SSSR, 137:92 (1961)
36. D. R. Inglis and E. Teller, Astrophys. J., 90:439 (1939).
37. W. Bottichen, Z. Phys., 150:336 (1957).
38. Lokhte-Khol'treven, Progr. Phys., 21:312 (1958).
39. W. L. Wiese and P. W. Murphy, Phys. Rev., 131:2108 (1963).
40. W. Botticher, et al., Z. Phys., 175:480 (1963).
41. P. M. Stone and L. Agnew, Phys. Rev., 127:1157 (1962).
42. H. R. Griem, Phys. Rev., 128:515 (1962).
43. K. R. Barnett, Atomnaya Tekhnika za Rubezhom (Atomic Science Abroad), No. 10, 31 (1958).
44. S. Yu. Luk'yanov and V. I. Kogan, Zh. Eksp. Teor. Fiz., 34:849 (1958) [Sov. Phys. − JETP, 7(1):587 (1958)].
45. M. M. But-slov, et al., Uspekhi Nauchnoi Fotografii (Prog. Sci. Photography), 6:85 (1959).
46. S. P. Zagorodnikov, G. E. Smolkin, and G. V. Sholin, Zh. Eksp. Teor. Fiz., 45:1850 (1963) [Sov. Phys. − JETP, 18(5):1268 (1964)].
47. A. G. Plakhov, et al., Opt. i Spekt., 16:329 (1964).
48. E. K. Zavoiskii and S. D. Fanchenko, Optics, 4:1155 (1965).
49. Cascade Image Converters and Their Application; (translated from the English), Mir, Moscow, 1965.
50. M. M. But-slov, V. S. Komel'kov, and Yu. E. Nesterikhin, Uspekhi Nauchnoi Fotografii, 9:72 (1964).
51. C. Breton, in: International Conference on Ionized Phenomena in Gases, Vol. 5, Munich, 1961, p. 1913.
52. G. S. Spillman, et al., Appl. Optics, 2:205 (1963).

53. V. G. Koloshnikov, et al., Opt. i Spekt., 11:556 (1961).

54. V. G. Koloshnikov, et al., Abstracts of Reports from the XIII Conference on Spectroscopy, Izd-nie LGU, Leningrad, 1961.

55. V. G. Koloshnikov and S. L. Mandel'shtam, Nucl. Fusion, Suppl. III, 1232 (1962).

56. S. Tolansky, High Resolution Spectroscopy, Methuen, London, 1955.

Chapter 5

Continuous Spectra. Determination of

Electron Density and Temperature

§ 5.1. Bremsstrahlung and Recombination
Radiation in the X-Ray, Ultraviolet,
and Visible Regions of the Spectrum

The shape and intensity of a continuous radiation spectrum from a plasma is determined by the interplay of the following processes:

1. Bremsstrahlung produced by the interaction of an electron with ions (free−free transitions).

2. Recombination radiation produced by radiative capture of an electron by an ion (free−bound transitions).

3. Bremsstrahlung produced as a result of electron−electron interactions.

The reaction rate for each of these processes is a function of the temperature and density. However, the relative role of each of the processes can vary for different spectral regions and for different electron temperatures. The variation of the coefficients and the complicated nature of the dependence of the radiation intensity on electron temperature means that under certain conditions the intensity of the radiation in some spectral range can be a rather weak function of the temperature; on the other hand, the absolute intensity of that portion of the spectrum can, with a high degree of accuracy, be used to determine the density, provided even relatively limited information is available on the temperature. Also the exponential nature of the dependence on the radiation frequency and temperature makes it possible to determine the electron temperature from the shape of the short-wave region of the spectrum even when plasma density is not known.

Of the three processes listed above as being responsible for the intensity of a continuous spectrum, we shall be interested only in the first two. The last process can play an important role only in a plasma with relativistic electrons; it does not appear that this process will be useful for diagnostic purposes in the near future.

The recombination spectrum exhibits appreciable intensity near the series edges. It reaches the edge of a series ν_0 from the short-wave side, increasing exponentially as the frequency is reduced to the value $\nu = \nu_0$. The spectral density of the radiation due to recombination for a level with a principal quantum number κ per unit volume of plasma (Maxwellian electron velocity distribution) is given by the relation

$$I_{\text{rec}} \, d\nu = A n_i n_e \left(\frac{E_H}{kT_e}\right)^{3/2} \left(\frac{E_{ik}}{E_H}\right)^2 \frac{g_{fb}}{\kappa} \, \xi_k \, e^{\frac{E_{ik}-h\nu}{kT_e}} \, d\nu. \tag{5.1}$$

Here $E_H = 13.6$ eV is the ionization energy for hydrogen; E_{ik} is the ionization energy for the shell for which the principal quantum number is κ; g_{fb} is the Gaunt factor for free−bound transitions [1, 2]; ξ_k is the number of vacant sites in the κ shell (if the shell is simple then $\xi_k = 2\kappa^2$) and finally, A is a numerical factor equal to $7 \cdot 10^{-40}$ erg · cm^3. By expressing E_{ik} in terms of κ and Z it is easy to show that the power in recombination radiation is proportional to Z^4/κ^3, that is to say, the recombination occurs primarily to the lower level and the power of the recombination continuum increases rapidly with ion charge.

The bremsstrahlung spectrum in a plasma is due to free−free transitions, in contrast with recombination radiation, and increases without limit in the long-wave direction. The power of the radiation per unit volume of plasma in a solid angle 4π in a frequency interval $d\nu$ is

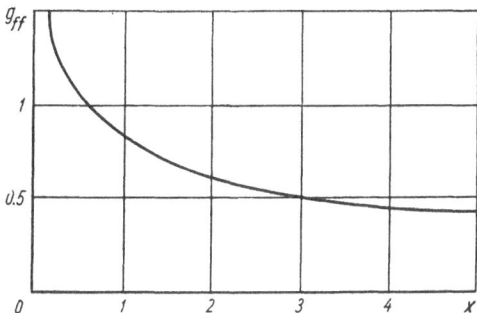

Fig. 5.1. Curve of the function g(hν/kT).

$$I_{\text{beam}} dv = AZ^2 n_i n_e \left(\frac{E_H}{kT_e}\right)^{1/2} e^{-\frac{h\nu}{kT_e}} g_{ff}\, dv, \qquad (5.2)$$

where $g_{ff}(h\nu/kT_e)$ is the Gaunt factor for free−free transitions
[3]. As in Eq. (5.1), $A = 7 \cdot 10^{-40}$ erg · cm^3. The Gaunt factor is a
weak function of its argument over a wide range of values and is
rather complicated. As an approximation we can write $g_{ff} \sim 1$;
a curve of this function is given in Fig. 5.1.

The form of Eq. (5.2) is valid for relatively low electron
temperatures (as long as the mean electron energy does not fall
in the relativistic range). In Eqs. (5.1) and (5.2) the quantity T_e
is given in degrees while E and hν are given in ergs. In the short-
wave region of the spectrum ($h\nu \gg kT_e$) the shape of the spectral
distribution both for free−free as well as free−bound transitions
is determined primarily by the exponential factor. When $h\nu \gg kT_e$
the logarithm of the spectral distribution of a continuous spectrum is essentially a straight line whose slope is given by
the electron temperature. This method of determining the
electron temperature is used widely in all experiments in which
x-ray radiation due to bremsstrahlung of thermal electrons is
observed [4, 5]. The method of determining T_e from the short-
wave tail of the x-ray spectrum is one of the most popular and reliable. The only complicating feature is the need for having reliable information on which part of the electron density is responsible for the temperature. The point here is that the electron
temperature, as determined from the shape of the continuous spectrum, may refer only to a small part of the plasma being studied.

In principle, cases can arise in which the distribution of particles is rather complicated. In particular, the plasma can sometimes be characterized by two electron temperatures: the electron temperature associated with a so-called hot component, and the electron temperature associated with a cold component of the plasma. In experiments on plasma heating by unstable electron beams [6] the density of hot plasma is determined by measuring the absolute yield of x-ray radiation from the volume of the system.

The total bremsstrahlung energy from a plasma per cubic centimeter (in ergs per second) is

$$Q_{\text{beam}} = 1.5 \cdot 10^{-25} \, Z^2 \, n_e \, n_i \, T^{\frac{1}{2}} \text{ eV.} \qquad (5.3)$$

For purposes of comparison we note that the intensity of the recombination radiation is given by

$$Q_{\text{rec}} = 5 \cdot 10^{-24} \, Z^4 \, n_e \, n_i \, T^{-\frac{1}{2}} \text{ eV.} \qquad (5.4)$$

From a comparison of these expressions it is evident that in a hydrogen plasma the intensity of the free−free radiation and the energy of the free−bound radiation are comparable at an electron temperature of about 30 eV. In a hydrogen plasma at temperatures above 100 eV the basic contribution to the radiation comes from the bremsstrahlung.

In a hydrogen plasma at a density of $\sim 10^{16}$ cm^{-3} or higher the intensity of the continuous spectrum can be measured in other regions beside the x-ray region, for example, the visible region and the ultraviolet region [7-9]. It has been shown in work by S. Yu. Luk'yanov and V. I. Kogan [7, 8] that the plasma density can be determined from the absolute intensity of the continuum radiation within a narrow spectral region in the visible portion of the spectrum. This method is useful in that the density can be determined under conditions for which the exact value of the electron temperature is not known. The idea of the method is based on the weak dependence of the spectral density of the continuous spectrum on temperature when $h\nu < kT_e$. Thus, for a change of electron temperature from 30 to 300 eV the intensity of the portion of the continuous spectrum that lies within the red changes by only 20%

if the plasma density remains fixed. This method has been used to determine the plasma density in intense pulsed discharges at times of peak compression. The intensity is observed at peak compression through flashes in the continuum, these being sufficient to carry out absolute measurements of the bremsstrahlung radiation in the visible. The density measured in this way was found to be 10^{17} cm^{-3}.

A series of refined measurements of the soft x-ray radiation spectrum for determining the electron temperature $T_e \sim 10^3$ eV has been carried out in the Sylla device. In these measurements the energy of the x-ray radiation in a given spectral range was determined by differential filters. The filters were thin organic films. The method of measurement and the results are described in detail in [10]. In addition to using organic filters for determining electron temperature from spectral analysis in the soft x-ray region, wide use is made of foils fabricated from aluminum and beryllium.

The x-ray radiation is emitted from the discharge chamber through a window made from beryllium foil and is then detected by a crystal and photomultiplier. The intensity of the radiation is changed after passing through filters of different thickness and it is assumed that in the energy range being investigated the pulse at the output of the photomultiplier is proportional to the photon energy. The dependence of intensity on the thickness of the beryllium, aluminum, and nickel filters is then compared with calculated curves. The values of the x-ray attenuation coefficients in a plasma with a Maxwellian velocity distribution is given in Table 5.1. In Fig. 5.2 we give absorption curves for aluminum filters as computed for various electron temperatures. The experimental points fit a curve that corresponds to an electron temperature of approximately 240 eV. Similar results are obtained with beryllium filters. It is found that in experiments with nickel filters the experimental curve is in much poorer agreement with the theoretical calculations. This discrepancy is evidently due to the strong absorption of radiation in nickel, which leads to the need for operating in a range of variation of the radiation intensity which extends over three orders of magnitude. It is completely reasonable that the accuracy of the relative measurements of the intensities differing by an order of magnitude may not be very high.

TABLE 5.1. Fraction of the Intensity of Bremsstrah-

Ab-sorber	Thick-ness, mg/cm²	T_e,			
		100	200	300	400
CH₂	0	1	1	1	1
	0.5	$2.02 \cdot 10^{-2}$	$4.27 \cdot 10^{-2}$	$7.14 \cdot 10^{-2}$	$1.08 \cdot 10^{-1}$
	1.0	$4.04 \cdot 10^{-3}$	$1.06 \cdot 10^{-2}$	$2.56 \cdot 10^{-2}$	$5.03 \cdot 10^{-2}$
	1.5	$9.98 \cdot 10^{-4}$	$3.61 \cdot 10^{-3}$	$1.30 \cdot 10^{-2}$	$3.10 \cdot 10^{-2}$
	3	$2.38 \cdot 10^{-5}$	$4.97 \cdot 10^{-4}$	$4.12 \cdot 10^{-3}$	$1.32 \cdot 10^{-2}$
	5	$3.91 \cdot 10^{-7}$	$1.38 \cdot 10^{-4}$	$1.70 \cdot 10^{-3}$	$6.60 \cdot 10^{-3}$
	7	—	—	—	—
	9	$1.21 \cdot 10^{-8}$	$2.80 \cdot 10^{-5}$	$5.37 \cdot 10^{-4}$	$2.63 \cdot 10^{-3}$
	11	—	$1.54 \cdot 10^{-5}$	—	$1.85 \cdot 10^{-3}$
	13	—	$9.16 \cdot 10^{-6}$	$2.37 \cdot 10^{-4}$	$1.37 \cdot 10^{-3}$
	15	—	$5.71 \cdot 10^{-6}$	$1.68 \cdot 10^{-4}$	$1.04 \cdot 10^{-3}$
	20	—	$2.12 \cdot 10^{-6}$	$8.18 \cdot 10^{-5}$	$5.85 \cdot 10^{-4}$
	30	—	$4.59 \cdot 10^{-7}$	$2.67 \cdot 10^{-5}$	$2.39 \cdot 10^{-4}$
	50	—	$5.24 \cdot 10^{-8}$	$5.47 \cdot 10^{-6}$	$6.73 \cdot 10^{-5}$
Be	0	1	1	1	1
	0.5	$2.13 \cdot 10^{-3}$	$3.21 \cdot 10^{-2}$	$8.88 \cdot 10^{-2}$	$1.51 \cdot 10^{-1}$
	1.0	$5.44 \cdot 10^{-4}$	$1.47 \cdot 10^{-2}$	$5.07 \cdot 10^{-2}$	$9.73 \cdot 10^{-2}$
	1.5	$3.53 \cdot 10^{-4}$	$8.63 \cdot 10^{-3}$	$3.46 \cdot 10^{-2}$	$7.19 \cdot 10^{-2}$
	3	$3.55 \cdot 10^{-5}$	$3.02 \cdot 10^{-3}$	$1.62 \cdot 10^{-2}$	$3.95 \cdot 10^{-2}$
	5	$7.47 \cdot 10^{-6}$	$1.22 \cdot 10^{-3}$	$9.43 \cdot 10^{-3}$	$2.35 \cdot 10^{-2}$
	7	$2.38 \cdot 10^{-6}$	$6.28 \cdot 10^{-4}$	$5.19 \cdot 10^{-3}$	$1.60 \cdot 10^{-2}$

lung from a Plasma Transmitted through an Absorber

eV					
500	600	800	1000	1200	1400
1	1	1	1	1	1
$1.52 \cdot 10^{-1}$	$1.69 \cdot 10^{-1}$	$2.53 \cdot 10^{-1}$	$3.25 \cdot 10^{-1}$	$3.90 \cdot 10^{-1}$	$4.41 \cdot 10^{-1}$
$8.20 \cdot 10^{-2}$	$1.11 \cdot 10^{-1}$	$1.81 \cdot 10^{-1}$	$2.47 \cdot 10^{-1}$	$3.09 \cdot 10^{-1}$	$3.60 \cdot 10^{-1}$
$5.61 \cdot 10^{-2}$	—	—	—	—	—
$2.80 \cdot 10^{-2}$	$4.69 \cdot 10^{-2}$	$9.23 \cdot 10^{-2}$	$1.41 \cdot 10^{-1}$	$1.90 \cdot 10^{-1}$	$2.36 \cdot 10^{-1}$
$1.57 \cdot 10^{-2}$	$2.85 \cdot 10^{-2}$	$6.21 \cdot 10^{-2}$	$1.02 \cdot 10^{-1}$	$1.44 \cdot 10^{-1}$	$1.85 \cdot 10^{-1}$
—	$1.97 \cdot 10^{-2}$	$4.64 \cdot 10^{-2}$	$7.98 \cdot 10^{-2}$	$1.17 \cdot 10^{-1}$	$1.54 \cdot 10^{-1}$
$7.27 \cdot 10^{-3}$	$1.46 \cdot 10^{-2}$	—	—	—	—
—	$1.14 \cdot 10^{-2}$	$3.00 \cdot 10^{-2}$	$5.55 \cdot 10^{-2}$	$8.56 \cdot 10^{-2}$	$1.17 \cdot 10^{-1}$
$4.21 \cdot 10^{-3}$	$9.10 \cdot 10^{-3}$	—	—	—	—
$3.35 \cdot 10^{-3}$	$7.54 \cdot 10^{-3}$	—	—	—	—
$2.07 \cdot 10^{-3}$	$4.93 \cdot 10^{-3}$	—	—	—	—
$9.75 \cdot 10^{-4}$	$2.75 \cdot 10^{-3}$	—	—	—	—
$3.36 \cdot 10^{-4}$	$1.02 \cdot 10^{-3}$	—	—	—	—
1	1	1	1	1	1
$2.15 \cdot 10^{-1}$	$2.70 \cdot 10^{-1}$	—	—	—	—
$1.48 \cdot 10^{-1}$	$1.97 \cdot 10^{-1}$	$2.85 \cdot 10^{-1}$	$3.59 \cdot 10^{-1}$	$4.24 \cdot 10^{-1}$	$4.76 \cdot 10^{-1}$
$1.16 \cdot 10^{-1}$	$1.59 \cdot 10^{-1}$	—	—	—	—
$7.02 \cdot 10^{-2}$	$1.04 \cdot 10^{-1}$	$1.72 \cdot 10^{-1}$	$2.37 \cdot 10^{-1}$	$2.97 \cdot 10^{-1}$	$3.49 \cdot 10^{-1}$
$4.56 \cdot 10^{-2}$	$7.14 \cdot 10^{-2}$	$1.28 \cdot 10^{-1}$	$1.86 \cdot 10^{-1}$	$2.41 \cdot 10^{-1}$	$2.91 \cdot 10^{-1}$
$3.30 \cdot 10^{-2}$	$5.42 \cdot 10^{-2}$	$1.03 \cdot 10^{-1}$	$1.55 \cdot 10^{-1}$	$2.07 \cdot 10^{-1}$	$2.54 \cdot 10^{-1}$

TABLE 5.1

Absorber	Thickness, mg/cm^2	T_e, 100	200	300	400
Al	9	$9.45 \cdot 10^{-7}$	$3.68 \cdot 10^{-4}$	$3.52 \cdot 10^{-3}$	$1.17 \cdot 10^{-2}$
	11	—	$2.33 \cdot 10^{-4}$	—	$9.05 \cdot 10^{-3}$
	15	$1.19 \cdot 10^{-7}$	$1.10 \cdot 10^{-4}$	$1.46 \cdot 10^{-3}$	$5.85 \cdot 10^{-3}$
	0	1	1	1	1
	0.5	$3.00 \cdot 10^{-3}$	$2.77 \cdot 10^{-2}$	$7.09 \cdot 10^{-2}$	$1.15 \cdot 10^{-1}$
	1.0	$3.20 \cdot 10^{-4}$	$8.72 \cdot 10^{-3}$	$3.05 \cdot 10^{-2}$	$5.66 \cdot 10^{-2}$
	1.5	$8.53 \cdot 10^{-5}$	$4.36 \cdot 10^{-3}$	$1.79 \cdot 10^{-2}$	$3.52 \cdot 10^{-2}$
	3	$9.38 \cdot 10^{-6}$	$1.17 \cdot 10^{-3}$	$5.98 \cdot 10^{-3}$	$1.29 \cdot 10^{-2}$
	5	$1.57 \cdot 10^{-6}$	$3.47 \cdot 10^{-4}$	$2.04 \cdot 10^{-3}$	$4.69 \cdot 10^{-3}$
	7	$4.13 \cdot 10^{-7}$	$1.30 \cdot 10^{-4}$	$8.36 \cdot 10^{-4}$	$1.98 \cdot 10^{-3}$
	9	$1.36 \cdot 10^{-7}$	$5.44 \cdot 10^{-5}$	$3.71 \cdot 10^{-4}$	$9.01 \cdot 10^{-4}$
	11	—	$2.45 \cdot 10^{-5}$	—	$4.29 \cdot 10^{-4}$
	15	$9.35 \cdot 10^{-9}$	$5.61 \cdot 10^{-6}$	$4.20 \cdot 10^{-5}$	$1.06 \cdot 10^{-4}$
Ni	0	1	1	1	1
	0.5	$1.42 \cdot 10^{-3}$	$1.57 \cdot 10^{-2}$	$3.50 \cdot 10^{-2}$	$5.34 \cdot 10^{-2}$
	1.0	$2.49 \cdot 10^{-4}$	$3.08 \cdot 10^{-3}$	$7.61 \cdot 10^{-3}$	$1.31 \cdot 10^{-2}$
	1.5	$4.35 \cdot 10^{-5}$	$7.79 \cdot 10^{-4}$	$2.13 \cdot 10^{-3}$	$4.35 \cdot 10^{-3}$
	3	$8.98 \cdot 10^{-7}$	$2.19 \cdot 10^{-5}$	$1.08 \cdot 10^{-4}$	$4.75 \cdot 10^{-4}$
	5	$1.01 \cdot 10^{-8}$	$4.30 \cdot 10^{-7}$	$1.08 \cdot 10^{-5}$	$9.46 \cdot 10^{-5}$
	7	$1.52 \cdot 10^{-10}$	$3.25 \cdot 10^{-8}$	$2.59 \cdot 10^{-6}$	$3.14 \cdot 10^{-5}$

(continued)

eV					
500	600	800	1000	1200	1400
$2.55 \cdot 10^{-2}$	$4.34 \cdot 10^{-2}$	$8.65 \cdot 10^{-2}$	$1.34 \cdot 10^{-1}$	$1.82 \cdot 10^{-1}$	$2.28 \cdot 10^{-1}$
—	$3.59 \cdot 10^{-2}$	$7.48 \cdot 10^{-2}$	$1.18 \cdot 10^{-1}$	$1.64 \cdot 10^{-1}$	$2.07 \cdot 10^{-1}$
$1.43 \cdot 10^{-2}$	$2.62 \cdot 10^{-2}$	—	—	—	—
1	1	1	1	1	1
$1.57 \cdot 10^{-1}$	$1.81 \cdot 10^{-1}$	$2.38 \cdot 10^{-1}$	$2.87 \cdot 10^{-1}$	$3.30 \cdot 10^{-1}$	—
$8.28 \cdot 10^{-2}$	$1.04 \cdot 10^{-1}$	$1.43 \cdot 10^{-1}$	$1.77 \cdot 10^{-1}$	$2.10 \cdot 10^{-1}$	$2.40 \cdot 10^{-1}$
$5.29 \cdot 10^{-2}$	$6.83 \cdot 10^{-2}$	—	—	—	—
$2.02 \cdot 10^{-2}$	$2.70 \cdot 10^{-2}$	$3.99 \cdot 10^{-2}$	$5.36 \cdot 10^{-2}$	$6.97 \cdot 10^{-2}$	$8.67 \cdot 10^{-2}$
$7.61 \cdot 10^{-3}$	$1.04 \cdot 10^{-2}$	$1.63 \cdot 10^{-2}$	$2.35 \cdot 10^{-2}$	$3.31 \cdot 10^{-2}$	$4.46 \cdot 10^{-2}$
$3.28 \cdot 10^{-3}$	$4.57 \cdot 10^{-3}$	$7.55 \cdot 10^{-3}$	$1.20 \cdot 10^{-2}$	$1.85 \cdot 10^{-2}$	$2.69 \cdot 10^{-2}$
$1.52 \cdot 10^{-3}$	$2.16 \cdot 10^{-3}$	$3.85 \cdot 10^{-3}$	$6.76 \cdot 10^{-3}$	$1.15 \cdot 10^{-2}$	$1.81 \cdot 10^{-2}$
—	$1.07 \cdot 10^{-3}$	$2.09 \cdot 10^{-3}$	$4.13 \cdot 10^{-3}$	$7.71 \cdot 10^{-3}$	$1.30 \cdot 10^{-2}$
$1.88 \cdot 10^{-4}$	$2.92 \cdot 10^{-4}$	—	—	—	—
1	1	1	1	1	1
$7.31 \cdot 10^{-2}$	$9.33 \cdot 10^{-2}$	$1.35 \cdot 10^{-1}$	$1.78 \cdot 10^{-1}$	$2.22 \cdot 10^{-1}$	$2.61 \cdot 10^{-1}$
$2.08 \cdot 10^{-2}$	$3.05 \cdot 10^{-2}$	$5.57 \cdot 10^{-2}$	$8.61 \cdot 10^{-2}$	$1.20 \cdot 10^{-1}$	$1.53 \cdot 10^{-1}$
$8.28 \cdot 10^{-3}$	$1.40 \cdot 10^{-2}$	$3.11 \cdot 10^{-2}$	$5.38 \cdot 10^{-2}$	$8.03 \cdot 10^{-2}$	$1.08 \cdot 10^{-1}$
$1.49 \cdot 10^{-3}$	$3.46 \cdot 10^{-3}$	$1.08 \cdot 10^{-2}$	$2.27 \cdot 10^{-2}$	$3.82 \cdot 10^{-2}$	$5.61 \cdot 10^{-2}$
$4.11 \cdot 10^{-4}$	$1.14 \cdot 10^{-3}$	$4.50 \cdot 10^{-3}$	$1.08 \cdot 10^{-2}$	$2.01 \cdot 10^{-2}$	$3.16 \cdot 10^{-2}$
$1.61 \cdot 10^{-4}$	$5.07 \cdot 10^{-4}$	$2.35 \cdot 10^{-3}$	$6.30 \cdot 10^{-3}$	$1.25 \cdot 10^{-2}$	$2.05 \cdot 10^{-2}$

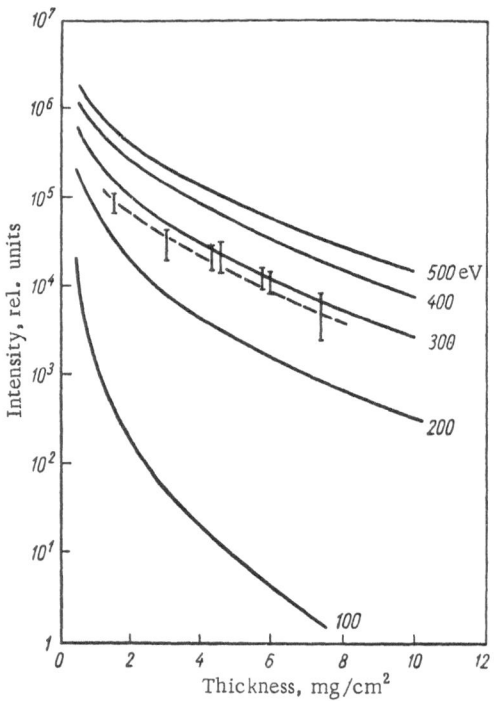

Fig. 5.2. The emission intensity in the continuous
spectrum as a function of the thickness of the aluminum filter.

The filter method is sometimes used for the analysis of harder
radiation from a plasma but the accuracy of such measurements
is rather low. The basic difficulty in the measurement of hard
x-ray radiation by the filter method is due to the appreciable
background of scattered radiation; it is not always easy to reduce
this background to a satisfactory level.

A superficial acquaintance with x-ray measurements frequently leads to the erroneous conclusion that it is a straightforward matter to interpret data obtained by simple measurements.
Actually, however, a unique analysis of the results can be carried
out only if one has available sufficiently precise data on the shape
of the spectrum and the radiation energy, in absolute units. In
determining the electron temperature from the mean energy of

the photons from the discharge (usually these measurements are
carried out by the filter method) it is easy to be in error by an
order of magnitude if one neglects the contribution of recombination radiation due to impurities which can be contained in the
plasma in substantial amounts.

As an actual example of an erroneous interpretation of results
we consider the measurement of density and electron temperature
in a hydrogen plasma which contains the impurity, oxygen, in the
amount of approximately 3% in terms of the number of atoms. We
shall now assume that at some initial time the density of this
plasma is approximately 10^{18} cm^{-3}, the ion temperature is approximately 500 eV, and that the electron temperature is appreciably lower than the ion temperature. We also assume that
the time during which the plasma is confined by the external field
is approximately 1 μsec. These values of density and temperature
are typical of the initial stages in experiments on compression of
an intense pulsed discharge.

The change in electron and ion temperatures during the course
of the entire compression state is determined by the two following
basic processes: a) transfer of energy from the ions to the electrons as a result of Coulomb collisions, and b) loss of energy of
the electrons due to ionization and excitation of atoms. A balance
equation [11] that takes account of Coulomb collisions and line radiation is

$$\frac{dT_e}{dt} = \frac{4}{3}\sqrt{2\pi m}\frac{e^2 Ln}{\left(kT_e + \frac{m}{M}kT_i\right)} - \frac{2}{3k}\sum_{s,\,r}\frac{n_r^*}{n_e}h\nu_{r,\,s}A_{r,\,s}, \qquad (5.5)$$

where m and M are the masses of the electron and proton; L is
the Coulomb logarithm (L $\sim$ 15); n_r^* is the density of impurity ions,
these being in the r-th excited state; $A_{r,s}$ is the probability for a spontaneous transition from the state r to the state s; k is the Boltzmann constant. Estimates made on the basis of the limiting cases
of the populations in the levels show that the electron temperature
reaches its maximum value, close to $T_{i0}/2$ in a time less than
10^{-7} sec; it falls to several tens of electron volts. Subsequently,
all of the oxygen atoms are converted into highly ionized ions and
approximately 10% of these become fully ionized. The power in the

recombination radiation of this plasma exceeds the power of the bremsstrahlung in the energy range 900-1000 eV by several orders of magnitude. In this case the recombination spectrum pertains to the limits of the series OVII and OVIII, 739 and 871 eV, respectively. Thus, a measurement of the energy as carried out by the filter method for filters fabricated from aluminum or beryllium with a thickness of several microns yields a mean photon energy close to 10^3 eV.

Let us assume that the electron temperature during compression is taken to be $T_e = 10^3$ eV rather than several tens of electron volts, as is actually the case; by measuring the radiation energy in the range 900-1000 eV in trying to determine the density from the expression in the bremsstrahlung radiation in a hydrogen plasma we would obtain a value exceeding 10^{18} cm^{-3}. As a result the electron temperature would be found to be higher than it should be by more than an order of magnitude; the density would also be too high. The results would be nice for an experimenter who was trying to obtain record values in a plasma; on the other hand, these would have very little in common with the parameters actually being investigated.

Thus, in summarizing this section we point out that an analysis of the x-ray spectrum can provide both the electron temperature and the electron density for electrons at the measured temperature. The determination of the density by the absolute yield of the x-ray radiation from a hydrogen plasma can lead to errors as large as an order of magnitude due to the presence of small amounts of impurity ions characterized by high atomic numbers. Still higher errors can arise if there is a possibility of loss of fast electrons to the walls with subsequent emission of x-ray photons.

§ 5.2. Infrared Region
of a Continuous Spectrum

From the point of view of plasma diagnostics great interest also attaches to long-wave infrared radiation. By measuring the energy distribution in a relatively small range of wavelength in this region, in principle it is possible to obtain both the plasma density and the electron temperature. For this purpose it is necessary to choose a portion of the spectrum within which there occurs a transition from the volume radiation associated with bremsstrahlung to radiation from the surface of the plasma. The latter situation corresponds to radiation of a blackbody and is realized in the frequency region in which the absorption coefficient becomes large enough. The power radiated by a blackbody is independent of the plasma density, being determined only by the temperature. Consequently, in determining the electron temperature of a plasma it is necessary to make a proper choice of the spectral range and to carry out absolute measurements. Furthermore, having carried out measurements of the radiation power of the plasma in the frequency region in which it is essentially transparent, and using the value of the electron temperature obtained in this way, it is not difficult to calculate the intensity of the bremsstrahlung and the density of the plasma.

If one takes account of the induced radiation the plasma absorption coefficient [12] is

$$K = 1.78 \cdot 10^{-2} g_{ff} \nu^{-2} T_e^{-3/2} n_e n_i \text{ cm}^{-1}. \tag{5.6}$$

Here g_{ff} is the Gaunt factor; T_e is the temperature in degrees; ν is the radiation frequency. This formula applies when $h\nu \ll kT$.

Substituting typical values of the plasma density for a discharge stabilized by a magnetic field ($n_e \sim 10^{15}$ cm^{-3}, $T_e \sim 10^5$ °K) in Eq. (5.6) we obtain an absorption coefficient equal to unity for a radiation wavelength of several millimeters. It will be evident from this estimate that the conditions approximating a blackbody are realized only at extremely low frequencies, that is to say, when $h\nu \ll kT$.

The continuum radiation power for a plane plasma layer of thickness d per unit surface in the direction normal to this surface ($h\nu \gg kT_e$) is

$$I d\nu = 5.5 \; 10^{-39} Z^2 g_{ff} T_e^{-1/2} n_e n_i \frac{1 - e^{-Kd}}{K} d\nu \text{ erg/(cm}^2 \cdot \text{sec} \cdot \text{sr}). \tag{5.7}$$

When $g_{ff} = 1$ we obtain the semiclassical approximation. For the low-frequency regions use can be made of the expressions computed in [13-15],

$$g_{ff} = 0.55 \left\{ \ln \left[\frac{2kT}{\pi Z e^2 \nu m^2} \right] - 1.44 \right\}, \tag{5.8}$$

or

$$g_{ff} = 1.27\left(3.38 + \log \frac{T}{n^{1/3}}\right).\tag{5.9}$$

In the far infrared, for a wide range of densities and temperatures in a laboratory plasma the factor g_{ff} is approximately 4.

Assuming that Kd ≫ 1 and substituting the value of K from Eq. (5.6) in Eq. (5.7), we can easily obtain an expression for the spectral density of the surface radiation (blackbody radiation), which is independent of density and is a linear function of the plasma temperature and a quadratic function of the radiation frequency. On the other hand, when Kd ≪ 1, that is to say for an optically thin plasma, in the frequency range $h\nu \ll kT_e$ the spectral density of the bremsstrahlung per unit volume of plasma is independent of frequency. Thus, the intensity of the bremsstrahlung radiation first increases in the direction of high frequencies going as ν^2; then, when the plasma becomes transparent [K(ν) = 0], it becomes frequency independent.

An investigation of the plasma emission spectrum in the region of millimeter and submillimeter waves for the purpose of determining the density and electron temperature was carried out on the Zeta device [16, 17]. The spectral resolution of the radiation was realized by means of a diffraction grating with constants of 0.5, 1.25, and 3.4 mm. The superposition of spectra corresponding to different orders was eliminated by the application of filters. (A description of the filters which can be used for investigating plasma radiation in the millimeter region is given in [18, 19].)

The radiation is detected by two detectors: a carbon bolometer [20] for recording the energy flux averaged over time, and a photoresistive element made of InSb, which is located in a magnetic field of 6000 Oe [21]. Both detectors work at liquid-helium temperatures. The resolution time of the second detector is of the order of microseconds, making it possible to determine the time behavior of the radiation intensity in various spectral ranges. The apparatus is calibrated by blackbody radiation.

The experimental curve showing the dependence of the radiation power on frequency exhibits a horizontal portion and then falls off at low frequencies. The spectrum is analyzed in accordance with Eqs. (5.6) and (5.7). The electron temperature is determined from the intensity of the radiation in the wavelength region 1-1.5 mm and it is possible to compute the density from measurements of the intensity in the range 0.2-0.3 mm. The analysis of the results yields the following values: $T_e \approx 10^5$°K and $n \approx 10^{15}$ cm^{-3}.

It should be noted that the method being described here for the determination of density and temperature is valid only if the measured portion of the spectrum lies far from any characteristic plasma frequency, since Eqs. (5.6) and (5.7) do not take account of any characteristic plasma oscillations. In experiments carried out by Harding and Roberts the plasma frequency was $\omega_0 = 3 \cdot 10^{11}$ sec^{-1}; consequently, the plasma oscillations could have lead to a distortion of the continuum spectrum and errors in the absorption coefficient.

If the effective characteristic plasma oscillations for a given case are not known it is difficult to state categorically that the values of n and T_e obtained in this way are accurate. The only verification of the applicability of Eqs. (5.6) and (5.7) under the conditions described by [17] is the good agreement between the shapes of the theoretical and experimental spectra; however, this verification is not altogether reliable. It should be remembered that the presence of a strong external magnetic field can reduce the effect of the plasma oscillations on the measured value of the absorption coefficient.

A much more convenient measurement, from the experimental point of view, is a measurement in the infrared region in the wavelength region of the order of microns [22]. For typical plasma parameters this radiation region is far from any characteristic plasma frequency.

Since the self-absorption of infrared radiation need not be taken into account over a wide range of density and electron temperature, the density of the plasma is related to the power in the continuum and the electron temperature by the formula

$$n = 2 \cdot 10^{19} \, T_e^{\frac{1}{4}} (\text{eV}) \left(\frac{\int\limits_{\Delta\nu} I(\nu) \, d\nu}{\Delta\nu g_{ff}} \right)^{\frac{1}{2}} \text{cm}^{-3}. \qquad (5.10)$$

Equation (5.10) is convenient in that it can be used to compute the density even if the value of T_e is known with poor accuracy. An error in the determination of the temperature as large as one order of magnitude leads to an error in the density which is smaller than a factor of two.

The detectors used for measuring the power of the radiation in absolute units and the time variation are usually fabricated from indium antimonide and germanium antimonide which are activated with gold and operate at liquid-nitrogen temperatures. These detectors can be used to measure densities of 10^{14} cm^{-3}.

The data given in the literature [23-26] for the spectral sensitivity of infrared detectors show wide discrepancies for wavelengths $\sim 1\,\mu$. These discrepancies are due to the fact that at wavelengths of the order of a micron the sensitivity of samples is determined primarily by the surface states. The region of wavelengths $\lambda < 1\,\mu$ can be easily be eliminated by the use of a mirror monochromator or by filters made from germanium or silicon. The first of these filters will stop radiation at wavelengths below $1.5\,\mu$, the short-wave limit of the second is $1\,\mu$. The infrared radiation from a vacuum chamber can easily be observed through a window made of lithium fluoride.

A vee-shaped molybdenum plate is used as a standard source of radiation when absolute measurements of infrared radiation are needed. Although the emittance of the molybdenum is rather small ($\sim 10\%$) the continuous radiation spectrum from a source of this shape with a small vertex angle is essentially the same as the radiation of a blackbody.

In order to estimate the emittance of a vee-shaped plate (this source is frequently called a wedge) we use the following formula:

$$\varepsilon = 1 - r^n, \tag{5.11}$$

where ε is the emittance power, that is to say, the ratio of the radiation of a given real source to the power of an absolute blackbody at the same temperature; $n \approx 180°/\gamma$ is the number of reflections experienced by a beam which enters the cavity before it is again reflected outward; γ is the vertex angle of the vee; r is the reflection coefficient; for molybdenum $r \sim 90\%$. Taking an angle $\gamma = 10°$, we find $\varepsilon = 0.85$. In cases in which one requires a high accuracy in the determination of the density it is necessary to make a curved vee-shaped source; the design of such a source is given in [27] together with experimental results.

The calibration of the source is reliable if the radiation of the blackbody is used to calibrate all of the measuring system at a frequency equal to the characteristic frequency of the change in the parameters of the plasma being investigated. Time modulation of the intensity of the source is easily realized by a rotating disk with an aperture located between the radiation source and the detector.

In the infrared region of the spectrum the basic error in measurements is associated with the background of line radiation corresponding to transitions between levels characterized by high principal quantum numbers. The contribution of the line radiation can be determined by photographing the spectrum. Photography of the spectrum is not required if it is possible to make estimates that show the intensity of the line infrared radiation to be negligibly small under the experimental conditions. A typical example of a plasma in which the intensity of the line radiation is small is the plasma in the leading edge of an electrodynamically accelerated plasmoid. Here, the line radiation arises only in the region of maximum intensity of the continuous spectrum [28].

In reviewing our analysis of methods based on treatment of the continuous spectrum of a plasma it is desirable to emphasize the following three points:

1. Absolute measurements of the spectral distribution and intensity of portions of a continuous spectrum over a wide range of wavelengths are of great interest for plasma diagnostics.

2. The spectral region most characteristic of a given state of a plasma is determined by the electron temperature.

3. The possibilities of the method are limited primarily by the background of line radiation and radiation from the surface of the wall of the vacuum chamber. Since the desired effect is proportional to n_e^2, while the noise level increases as n_e, measurements are possible starting at some value of the density determined by specific experimental conditions.

References

1. T. F. Stratton, in: Optical Spectrometric Measurement of High Temperatures (P. J. Dickerman, Ed.), The University of Chicago Press, Chicago, 1961, p. 99.
2. I. I. Sobel'man, Introduction to the Theory of Atomic Spectra [in Russian], Fizmatgiz, Moscow, 1963.
3. V. I. Kogan and A. B. Migdal, in: Plasma Physics and the Problem of a Controlled Thermonuclear Reaction, Pergamon, New York, 1958, Vol. 1.
4. W. B. Ard, et al., Phys. Rev. Lett., 10:87 (1963).
5. M. V. Babykin, et al., Zh. Eksp. Teor. Fiz., 47:1597 (1964) [Sov. Phys. — JETP, 20(1):1073 (1965)].
6. J. Allexeff, et al., Phys. Rev. Lett., 10:273 (1963).
7. V. I. Kogan, Plasma Physics and the Problem of a Controlled Thermonuclear Reaction, Pergamon, New York, 1958, Vol. 3.
8. S. Yu. Luk'yanov, Zh. Eksp. Teor. Fiz., 36:1621 (1959) [Sov. Phys. — JETP, 9:1155 (1959)].
9. H. B. Griem, A. C. Kolb, and W. R. Faust, Phys. Rev. Lett., 2:281 (1959).
10. F. C. Jahoda, et al., Phys. Rev., 119:843 (1960).
11. Yu. K. Zemtsov, V. D. Pis'mennyi, and I. M. Podgornyi, Dokl. Akad. Nauk SSSR, 155:312 (1964) [Sov. Phys. — Dokl., 9(3):219 (1964)].
12. C. W. Allen, Astrophysical Quantities, Izd. Inostr. Lit., Moscow, 1960.
13. L. Spitzer, Jr., Physics of Fully Ionized Gases, Interscience, New York, 1965.
14. P. A. Scheuer, Monthly Notices Roy. Astr. Soc., 120:231 (1960).
15. F. S. Smerd and K. C. Westford, Phil. Mag., 40:831 (1949).
16. G. N. Harding, et al., Proc. Phys. Soc., 77:1069 (1961).
17. G. N. Harding and V. Roberts, Nucl. Fusion, Suppl. III, 883 (1962).
18. V. Ya. Balakhanov and A. R. Striganov, Zh. Prikl. Spektr., 4:213 (1966).
19. V. Ya. Balakhanov, et al., Zh. Tekh. Fiz., 36:1383 (1966) [Sov. Phys. — Tech. Phys., 11(8):1032 (1967)].
20. W. S. Boyle and K. F. Podgers, J. Opt. Soc. America, 49:66 (1959).
21. E. H. Putey, Proc. Phys. Soc., 76:802 (1960).
22. L. A. Dushin, et al., Opt. i Spekt., 19:674 (1965).
23. P. Krause, MacGrauling, and MacKwistin, Infrared Radiation, Voengiz, Moscow (1964).
24. H. Fregerikse and R. Blunt, Proc. IRE, 43:1827 (1955).
25. D. H. Martin, Contemporary Physics, 4:139 (1962) and 4:187 (1963).
26. D. G. Avery, Proc. Phys. Soc., 67B:761 (1949).
27. S. A. Yakovlev, Proceedings of the 10th International Conference on Spectroscopy, Izd. LGU, L'vov, 1957, Vol. 1, p. 368.
28. G. G. Managadze, I. M. Podgornyi, and V. D. Rusanov, Zh. Tekh. Fiz., 37:2199 (1967) [Sov. Phys. — Tech. Phys., 12(12):1620 (1968)].

Chapter 6

Time-Resolved Photography

Information on the emission from a plasma is frequently
found to be of extremely great value in understanding the conditions of confinement of a plasma and the characteristics of its
interaction with a magnetic field. The compression of a plasma
under the effect of the discharge current and the development of
instabilities in an intense pulsed discharge were first observed by
streak photographs. This same technique was used to obtain the
first data on the interaction of fast plasmoids with a magnetic
field.

Depending on the parameters of the plasma, with all other conditions being equal the intensity of the emission in the visible
portion of the spectrum will be proportional to the density of electrons or the square of the density. A direct proportionality obtains between the electron density and the emission intensity when
the basic contribution is due to the line spectrum of neutral atoms,
assuming that the atomic density remains essentially unchanged
at the time of interest. At high densities, when the basic contribution to the radiation is in the continuous spectrum, the intensity is
proportional to the square of the density. If the line radiation of
the plasma is primarily emission from ionized atoms, the dependence on density is more complicated since the density of atoms
in a given state of ionization and the population of the excited levels
are themselves functions of the electron density. Thus, lacking
additional information, one is frequently led to an erroneous conclusion with respect to the spatial distribution of the density on the

basis of an analysis of photographs alone. For example, in photographing a jet of fully ionized hydrogen plasma of modest density it is not difficult to reach an incorrect conclusion on the basis of the data in a photograph if the concentration of neutral gas in the chamber is inhomogeneous; this is so since the radiation from a unit volume $h\nu n_0 n_e \langle \sigma v \rangle$ is proportional both to the electron density as well as the density of atoms of neutral gas.

The techniques employed in still photography and movie photography are usually not suitable for investigating the time behavior of a plasma. For example, in investigations of a pulsed high-power discharge the diameter of the plasma column may be approximately one centimeter. The time during which an appreciable change occurs in the geometry of the discharge at a compressional velocity of 10^7 cm/sec is approximately 10^{-6} sec. Consequently the exposure time cannot be greater than fractions of a microsecond; this time is several orders of magnitude smaller than the exposure time of conventional movie cameras.

One device that acts as a microsecond shutter is the Kerr cell [1, 2]. The principle of operation of this cell is based on the polarization of light passing through an anisotropic medium. In its simplest form a Kerr cell consists of two crossed polaroids between which there is a container (with plane-parallel transparent walls) filled with benzol. In the container there are two electrodes, the plane of orientation of these electrodes being parallel to the beam that passes through the polaroids. In the absence of an electric field the benzol is optically uniform and does not affect the polarization of the light that passes through it. In this state of the cell the intensity of the transmitted light is low and is determined primarily by the distance between inhomogeneities and between the crossed polaroids. In an electric field the benzol molecules become oriented along the field and the benzol becomes a birefringent medium. If the dimensions of the container and the intensity of the electric field are chosen in such a way that the phase shift of the ordinary and extraordinary rays is $\pi/2$, the polarization vector is rotated by 90°, corresponding to maximum transparency of the Kerr cell. If the path length in the benzol is 10-20 cm, rotation of the plane of polarization of visible light by 90° can be achieved through the use of an electric field of 30 kV/cm. Fields of this magnitude can be produced relatively easily in benzol.

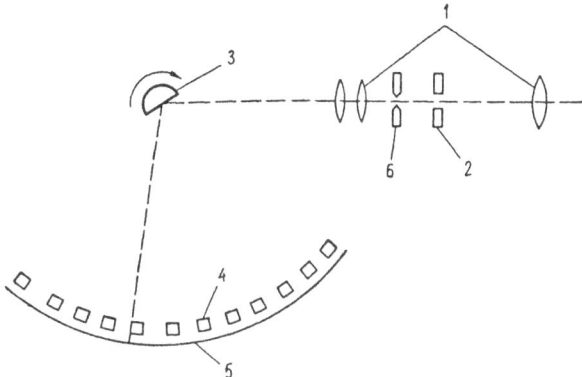

Fig. 6.1. Optical diagram for a rotating-mirror camera: 1) lenses; 2) diaphragm; 3) rotating mirror; 4) objectives; 5) film; 6) mirror.

Processing by high-voltage pulses serves to improve the electrical strength of benzol against breakdown.

The operation time of the Kerr cell is determined by the time required for reorientation of the molecules in the electric field and is approximately 10^{-9} sec. An exposure time of $\sim 10^{-9}$ sec is not possible for devices that use mechanical shutters. Although estimates show that the limiting achievable resolution of a system with a rotating mirror (which is limited by mechanical strength) is also 10^{-9} sec [3], the construction of such devices is essentially at the limits of modern-day technology. A basic disadvantage of the Kerr cell, when used as a fast shutter, is the fact that it is impossible (using a single camera) to take a series of pictures at a high repetition rate.

A series of successive frames with an exposure time of 0.4 μsec can be obtained by means of a high-speed camera with a rotating mirror. A schematic diagram of the principle of operation of the high-speed framing camera is shown in Fig. 6.1.

The image of the subject is formed by two objective lenses, between which is located an iris. Through the mirror and one of the many lenses located on the arc the image is transferred to the focal frame at which the film is located. When the mirror rotates the image of the object is successively projected through the lenses in the arc onto the film, in the form of individual fixed images. The fixed images are obtained by virtue of the fact that the light beam reflected by the mirror (image) is transferred from one lens to another with such a small angle of rotation that the rotation smears the vertical edges of the image on the film by an amount that is smaller than the resolving power of the optical system.

A special triggering device provides the synchronizing pulse. The pulse triggers the device in which the plasma is being investigated at the exact time at which the rotating mirror is in a position such that the most interesting portion of the process will be recorded on the film. The constancy of the speed of rotation on the mirror is controlled by a pulse generator controlled by a quartz-crystal oscillator. The adjustment of the high-speed mirror camera is extremely simple and does not require any special training. Many interesting phenomena have been observed by means of this device. In particular, an analysis has been made of the instability in the shape of a plasma under various conditions [4, 5]. Typical photographs of the penetration of a plasma jet into a magnetic field and the subsequent confinement of the plasma in a system with a minimum B field are shown in Fig. 6.2. In this case the magnetic field is produced by two coils; the coil positions are easily determined from the photographs by the two dark regions intersecting the image of the plasma jet in the horizontal direction. When the coils are connected in opposition the magnetic field increases in all directions from the center. With this configuration the magnetic field vanishes at the center. The injection of the plasma occurs along the axis of symmetry of the magnetic field. A coaxial injector is used.

The currently available commercial mirror camera, the SFR, does not have a high-speed shutter which can expose the objective in a time less than one rotation of the mirror at a framing rate of $2 \cdot 10^6$ frames/sec. This shortcoming can be avoid by means of a mechanical or electrical shutter [7]. The first of these is a rotating disk with an aperture which is located in front of the objective. The rate of rotation of the disk and the injection of the discharge being studied are determined by means of a magnetic pickup of the same kind used for controlling the rotation of the mirror in the SFR. A second kind of shutter is based on the reduction of the transmission of plates of Plexiglas or silicate glass due to cracking

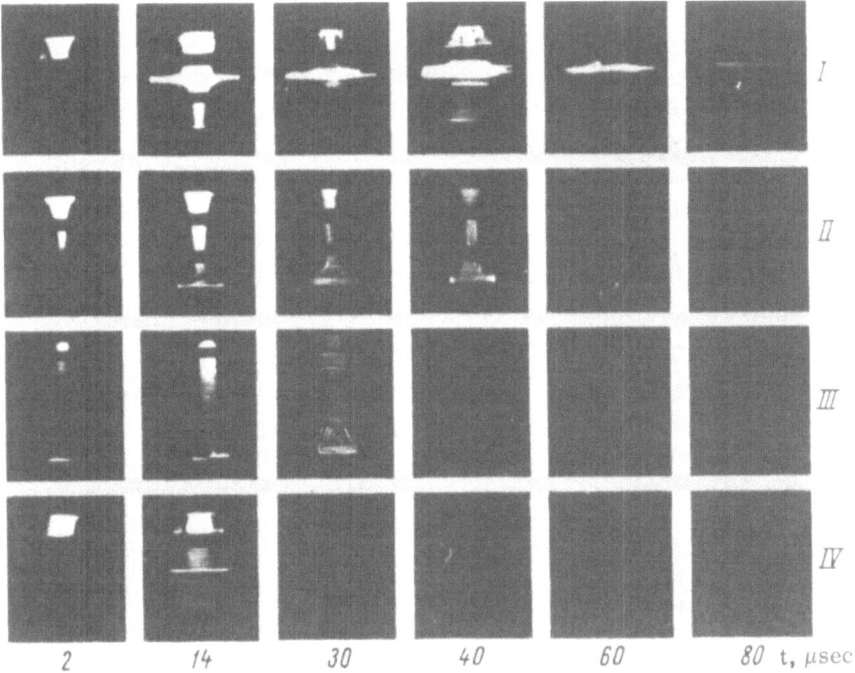

Fig. 6.2. Photograph showing the interaction of a plasmoid with magnetic fields of different configuration. The magnetic field is produced by two plane coils. 1) The coils are connected in opposition; 2) the coils are connected in the same direction and the field between them is approximately uniform; 3) the coils are shifted and connected in the same direction in a mirror configuration; 4) there is no magnetic field and the luminous horizontal band in the second frame is produced by plasma striking the edge of the flange.

caused by an exploding wire. The exploding wire loop is located between two glass plates. The convenience of this device lies in the fact that it can be triggered at any predetermined time. The principal shortcoming of the device is the fact that it must be replaced after each experiment.

In the investigation of processes characterized by axial symmetry sometimes it is convenient to replace the high-speed camera by an optical sweep system (streak camera). To illustrate the optical sweep method we consider an application in the observation of uniform compression of a plasma column. By isolating a beam of radiation from the plasma by a slit located perpendicularly to the axis of the plasma column and using the appropriate optics it is possible to focus the image of the slit on a film after reflection of the beam from a rotating mirror. The axis of rotation of the mirror is parallel to the slit. As the mirror rotates different portions of the film receive the projection of the image of the slit, corresponding to different instants of time. The streak camera has been used to observe the compression of a plasma column under the effect of its self-field (due to the discharge current) and the subsequent expansion; it was possible to observe the structure of the shock waves and the propagation of these waves to the axis of the discharge [6].

Lines in the visible portion of the spectrum have not been adequately investigated in experiments with high-temperature plasmas so that special interest attaches to the photography and streak photography of images in the ultraviolet and x-ray regions. The image of an object can be obtained through the use of a pinhole camera. In order to obtain sweep of the image of the slit in time, in this case it is convenient to make use of a rapidly rotating film [8].

Data obtained by streak photographs or framing photographs in high-speed cameras are extremely valuable if it is possible to relate the pictures to curves that illustrate the time behavior of various processes in the plasma. Synchronization can be achieved most simply by making time marks on the film or by recording, with an oscilloscope, the time at which the light pulse appears which exposes a given frame. The first method is applied both in the framing camera and in the streak camera. An extremely short voltage pulse is applied to a spark gap located in the field of view of the SFR. The emission from the spark is detected on the film and the voltage pulse is recorded by one of the beams of a dual beam oscilloscope. The second beam is used to display the curve showing the variation of whatever parameter is being investigated. A second method of synchronization is used in a framing camera [5]. In this case the radiation reaching the photographic film during the exposure time is determined by the frame and also strikes the photocathode of a photomultiplier being recorded by an oscilloscope. The marking on this frame is carried out by the subsequent illumination of the objective in the neighboring frame.

The information on the behavior of a plasma is much more valuable if the photography is carried out in spectrally dispersed light rather than in white light. A plasma in a discharge chamber will generally be inhomogeneous both with regard to density and temperature; frequently this situation is such that the basic contribution to the radiation in the visible region comes from lines of neutral or weakly ionized atoms which are located in a region in which the electron temperature is low, i.e., a region which is of little interest for the investigation. In high-temperature plasmas we find that the lines of the radiation are due to highly ionized atoms. Consequently, in order to obtain pictures of that portion of the plasma in which a high temperature obtains it is necessary to use a filter which cuts off the radiation from weakly ionized atoms. The most suitable filters for this purpose are interference filters, which can isolate regions several angstroms in width in the visible portion of the spectrum [9, 10]. Since filters for the required spectral range are not always available in the laboratory, it is also possible to isolate the radiation of highly ionized atoms through the use of a monochromator. However, in this case the light power of the optical system is much poorer.

In experiments with high-temperature plasmas different portions of the object very frequently acquire a shape which is approximately that of a circular cylinder with varying radius. In this case, as we have already pointed out, it is not necessary to obtain a photograph of the entire plasma object. Rather, it is sufficient to photograph the object through a narrow slit which is oriented perpendicularly to the plasma axis. Projecting the image of the objective on the slit of a spectrograph (the slit of the spectrograph is assumed to be perpendicular to the axis of symmetry of the plasma) we record on a film the nonuniformly blackened lines of radiation, each of which represents a photograph taken through the slit of the spectrograph. In other words we obtain a series of photographs through the slit, each photograph being taken by the light of its own spectral line. By using a spectrograph with quartz optics and working with a reasonably high-temperature plasma it is an easy matter to obtain an emission intensity which is uniformly distributed along a diameter for lines of multiply charged ions. For this purpose it is convenient to use the lines CI 6578, CII 4267, CIII 4647, and CIV 2530 Å which lie in an available spectral region; furthermore, these lines can be observed without the need for introducing impurities, since carbon is almost always present in any discharge chamber. Although photographs taken by this method give a frame averaged over time, the averaging occurs only over the time interval in which emission comes from a given multiply ionized ion. In other words, to some degree photography in the spectrally dispersed light replaces fast photography.

In a stationary plasma with a temperature gradient which is directed inward, the emission of multiply charged ions will be observed close to the axis while the lines of weakly ionized ions will be emitted primarily from the edge. Consequently, even without going through a complicated analysis of the results it is possible to obtain information concerning the electron temperature.

In the investigation of low-density plasmas, the intensity of the plasma radiation is small and mechanical sweep devices become inconvenient because of their low light transmission. In this case, high-speed photographs have been taken with the so-called image converter. An important advantage of the image converter is its high resolution time, which is far better than that which can be achieved with mechanical sweeps. The principles of operation of this device have been described in §4.4 in the analysis of methods for time sweep of spectral lines.

What is evidently the first widespread application of the image converter in the time analysis of rapidly occurring processes appeared in 1949 when a device was developed [11, 12] known as the PIM (pulse converter). A diagram of the pulsed converter (PIM) is shown in Fig. 6.3. The gain of the electron-optical system in the PIM-3 and PIM-4 is two. The resolving power of the screen is 300 pairs of lines per centimeter. The electron beams that form the electron-optical image are deflected by deflection plates. Two pairs of deflection plates move the beam along orthogonal directions and serve for obtaining a two-dimensional sweep of the image. The two other pairs of plates comprise an element which provides a small exposure time, a balanced electrostatic shutter.

The principle of operation of the balanced shutter is the following: with no potential difference between plates P_1 and P_2 electrons which form the image move freely through the iris and form the image on the fluorescent screen. As the potential difference between these plates is increased the image is deflected. At some value of the field the electrons strike the iris and the image on the screen is quenched. Simultaneously, the image on

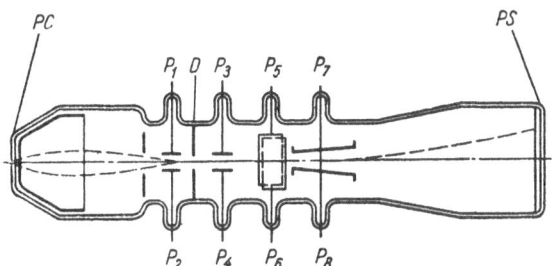

Fig. 6.3. Electron-optical image converter.

the screen is smeared out. The smearing of the image due to
the displacement in time of the cutoff can be reduced to a minimum by reducing the leading edge of the voltage pulse. In the
PIM more sophisticated methods are used which make it possible
to avoid the smearing of the image without the application of advanced pulsed techniques. The displacement of the image by plates
P_1 and P_2 is balanced by the electric field which is applied between plates P_3 and P_4, to which is applied the same difference of
potential but with opposite polarity. Plates P_5, P_6, P_7, and P_8 are
used to obtain the sweep of the image in time. By applying short
rectangular pulses to the plates it has been possible to obtain a
series of frames with exposure times of 10^{-8}-10^{-7} sec. This time
resolution is not the limiting value; according to available estimates, the minimum exposure time can be reduced to 10^{-12} sec.

Multiframe photography is realized by abrupt displacements
of the electron images in the horizontal and vertical directions.
The displacement of the image occurs by cutting off the electron-
optical shutter. The image is displaced by applying multistep
voltages to the plates. Each step corresponds to a single displacement of the frame. The number of frames in a row and the number
of rows is determined by the number of steps of the deflection
pulses while the repetition rate is determined by the length of these
pulses. The frequency of photographs of an optical chamber with
a PIM-3 obtained in [13] has been as high as $5 \cdot 10^6$ frames/sec
with an exposure of $5 \cdot 10^{-8}$ sec and a resolving power of 30 lines
per millimeter.

The usefulness of the EIC is not limited by its high resolution time and its high quantum sensitivity, as compared with photo-

graphic films (the quantum sensitivity of the electron-optical converter is ten times greater than that of an emulsion). Because of its high gain (10^4-10^5) the electron-optical image converter can be used to obtain information on the source of light under conditions in which a countable number of photons strikes the cathode. Although this information may not give all the features of the distribution of the brightness of the light source, the limitation of information lies in the source of radiation itself.

The image converter need not be limited to the visible region of the spectrum. For example, pictures can be obtained in infrared light by means of a device with a cesium oxide cathode shielded against visible light by a germanium filter. Photographs in x-ray light can be obtained through the use of fluorescent screens that are sensitive to x rays.

References

1. É. Angerer, Technique of Physical Experiments [in Russian], Fizmatgiz, Moscow, 1962, p. 384.
2. A. M. Zarem, F. R. Marshall, and F. P. Poole, Trans. AIEE, 68:84 (1949).
3. G. L. Sherman, Uspekhi Nauchnoi Fotografii (Progr. Sci. Photography), 6:93 (1959).
4. I. M. Podgornyi and V. N. Sumarokov, Nucl. Fusion, Suppl. I, 87 (1962).
5. N. A. Borzunov and D. V. Orlinskii, Atomnaya Énergiya, 4:149 (1958) [Sov. Atomic Energy, 4(2):195 (1958)].
6. V. S. Komel'kov, Zh. Eksp. Teor. Fiz., 35:16 (1958) [Sov. Phys. – JETP, 8(1):10 (1959)].
7. V. S. Komel'kov and V. V. Tserevitinov, Uspekhi Nauchnoi Fotografii (Progr. Sci. Photography), 9:184 (1964).
8. V. S. Vasilevskii, N. V. Krasnov, and V. S. Mukhovatov, Prib. Tekh. Éksp., 6:2(138) (1961).
9. G. V. Rozenberg, Usp. Fiz. Nauk, 47:3 (1952).
10. G. V. Rozenberg, Usp. Fiz. Nauk, 47:3 (1952) (sic).
11. E. K. Zavoiskii and S. D. Fanchenko, Appl. Optic, 4:1155 (1965).
12. M. M. Butslov, Uspekhi Nauchnoi Fotografii (Progr. Sci. Photography), 6:76 (1959).
13. M. I. Pergament, Yu. E. Nesterikhin, and V. S. Komel'kov, Uspekhi Nauchnoi Fotografii (Prog. Sci. Photography), 9:64 (1964).

Chapter 7

Determination of the Dielectric Constant

of a Plasma

§ 7.1. Measurement of Plasma

Density with Microwaves

The propagation of electromagnetic waves in a plasma is determined by the value of the dielectric constant. In the absence of a magnetic field the dielectric constant can be written

$$\varepsilon = 1 - \frac{4\pi e^2\, n}{m_e\, \omega^2} \cdot \frac{1}{1 - i\, \dfrac{\nu_{col}}{\omega}}, \tag{7.1}$$

where ω is the angular frequency of the electromagnetic wave; ν_{col} is the electron collision frequency in the plasma. The frequency of collisions with charged particles is given by

$$\nu_{col} = \frac{2\cdot 10^{-5}\, n_i\, Z}{T_e^{3/2}\ (eV)}.$$

We shall limit our analysis to that range of plasma densities and temperatures for which the frequency of the microwave radiation used for probing the plasma is much greater than the particle collision frequency. It will be evident that in practice this limitation does not really limit the application of the method since it does not limit the range of temperatures and densities which are

141

of most interest in the physics of hot plasmas. On the other hand, in this region of temperature and density, where ε is independent of the frequency of Coulomb collisions in the plasma, there is a unique relation between the dielectric constant and the density. The dielectric constant is given by the expression

$$\varepsilon = 1 - \frac{\omega_0^2}{\omega^2}. \tag{7.2}$$

Here ω_0 is the angular plasma frequency

$$\omega_0 = \sqrt{\frac{4\pi e^2 n}{m}}, \tag{7.3}$$

or

$$\omega_0 = 5.6 \cdot 10^4 \sqrt{n}.$$

Equation (7.2) is interesting in that when $\omega = \omega_0$ the refractive index $N = \sqrt{\varepsilon}$ vanishes; any further increase in n for a fixed value of the frequency means that the frequency of the electromagnetic wave becomes imaginary. An electromagnetic wave incident on the boundary of a plasma of sufficiently high density experiences total internal reflection and does not penetrate the plasma. Thus, if along the path of a microwave beam with a fixed wavelength there is a region occupied by a plasma whose density increases with time, when the condition $\omega_0 > \omega$ is satisfied the detector will no longer record radiation transmitted through the plasma. In this way we have the possibility of determining the density at the time at which the signal from the microwave generator is cut off. The density, more precisely, the maximum plasma density along the path of the beam at this time, is then given by

$$n_{\text{cr}} = \frac{\omega^2 m}{4\pi e^2} \quad \text{or} \quad n = \frac{1.1 \cdot 10^{13}}{\lambda^2}, \tag{7.4}$$

where λ is the wavelength of the microwave radiation in centimeters. This method of determining plasma density is called the cutoff method. In applying the cutoff method for measuring density it is necessary to recall that the idea of reflection of a wave from the boundary of a dense plasma is a gross simplification. Actually, the amplitude of the wave does not vanish at the interface between the plasma and vacuum. Rather, it decays exponentially in the depth of the plasma, the so-called anomalous skin depth.

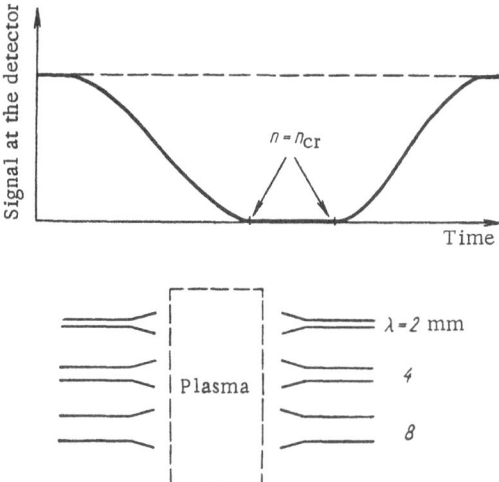

Fig. 7.1. The intensity of the microwave radiation
transmitted through a plasma. The dashed line
shows the intensity in the absence of plasma.

Hence, the determination of plasma density by the cutoff method
can be used reliably only when

$$l \gg \frac{c}{\omega_0},\qquad(7.5)$$

where l is the thickness of the plasma layer whose density is higher
than the critical value, while ω_0 is the plasma frequency.

The arrangement used for probing a plasma with centimeter
or millimeter radiation is extremely simple. The microwave signal at a fixed wavelength λ passes through the waveguide into a
horn, is transmitted through the plasma, and is then detected by a
detector furnished with a receiving horn. In detecting the signal
at the output of the detector we can obtain a curve (Fig. 7.1) which
contains points at which the density is computed from Eq. (7.4).
It is evident that there is a limit on the information concerning
plasma density which can be obtained by the microwave method.
The information is much more valuable if it is possible to use
several generators, each of which is tuned to a different wavelength; however, this complicates the experimental arrangement
to a considerable degree. In plasma research the most common
wavelengths are 2, 4, 8, and 30 mm. When the wavelength of the

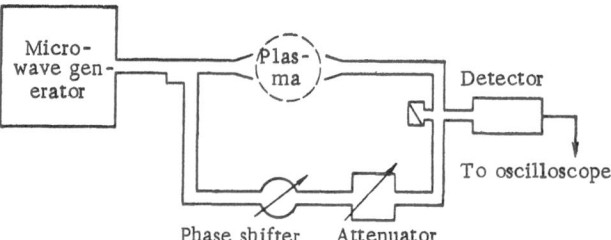

Fig. 7.2. A simple two-beam interferometer.

probing radiation is much smaller than the characteristic dimensions of the plasma being investigated it is possible to obtain additional information concerning the spatial distribution of the
plasma. In this case use is made of several probing beams from
one microwave generator. The requirement that the wavelength of
the microwave radiation be small compared with the dimensions
of the plasma is important not only in experiments on the determination of the spatial distribution of the density, but in all diagnostics experiments with microwave beams. In a plasma whose
characteristic dimensions are comparable with the wavelength
there is a high probability of encountering diffraction effects, in
which case complete cutoff of the microwave signal is not observed at any density.

Microwave probing of a plasma is also used in the determination of the density of a plasma in the uniform magnetic field. If
the electric vector of the probing microwave field is parallel to the
lines of force of the magnetic field, Eq. (7.2) remains valid for
the dielectric constant because with this orientation of the fields
the propagation of the electromagnetic wave in the plasma in the
presence of the external magnetic field is the same as with no
magnetic field.

More detailed information on plasma density and the variation of this quantity with time, together with the variation in dimensions of the plasma, can be obtained by observation of the
interference between two beams from the same generator. One
of the beams (the reference beam) goes to the detector directly
through a waveguide while the second is transmitted through the
plasma being investigated. A diagram of a simple interferometer
arrangement is shown in Fig. 7.2. The attenuation of one of the

beams by the attenuator is adjusted in such a way that the amplitude of both beams is the same in the absence of the plasma. Then a specified phase shift is set between the two beams in the absence of the plasma. A phase shift of π causes the signal at the detector to vanish.

We note that the phase shift can be controlled by a phase shifter in one of the branches of the bridge or by changing the distance between the horns. The introduction of plasma into the region beteen the horns causes an additional phase shift and causes an interference pattern between the two waves: the plasma causes an additional phase shift $\varphi(N)$ due to the change in the refractive index of the medium,

$$\left. \begin{aligned} E_1 &= E_0\,e^{i\,[\omega t + \varphi\,(N)]}, \\ E_2 &= E_0\,e^{i\,[\omega t + \pi]}. \end{aligned} \right\} \qquad (7.6)$$

The amplitude of the signal recorded by the detector will be modulated because in the general case the plasma density, and consequently $\varphi(N)$, are functions of time,

$$E = E_0\,(e^{i\varphi\,(N)} - 1)\,e^{i\omega t}. \qquad (7.7)$$

The pattern observed in the experiment differs from that described by Eq. (7.7) because as the density is increased the amplitude of the microwave beam that passes through the plasma is reduced and the modulation of the signal that appears at the detector exhibits a decaying pattern. A typical oscillogram obtained by means of an interferometer is shown in Fig. 7.3. Before the plasma appears there is no signal. As the plasma density increases a first interference peak is observed and this is followed by the others; when the density reaches the value given by Eq. (7.4) the transmission of the beam through the plasma is cut off and the oscilloscope shows a signal corresponding to the microwave radiation that is transmitted through the reference arm. As the density is reduced a pattern develops with the opposite sequence, with the difference that the time intervals between peaks become larger. This feature is a consequence of the slow decay of the plasma density.

We now consider the quantitative relation that obtains between the individual interference peaks and the density. When the refractive index of the medium changes the phase shift can be ex-

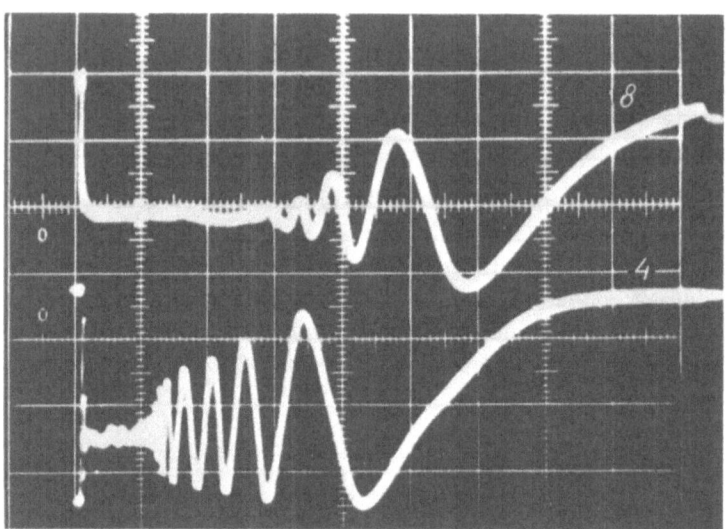

Fig. 7.3. Interference pattern obtained simultaneously at two wavelengths $\lambda = 8$ and $\lambda = 4$ mm. Because of the rapid increase in density up to the value n_{cr} the interference peaks are not resolved in the apparatus. The interference pattern is, however, resolved when the plasma decays.

pressed in the following way:

$$\varphi = \frac{2\pi}{\lambda} \int_0^l [N_0 - N(x)]\, dx, \qquad (7.8)$$

where l is the plasma dimension; $N = \sqrt{\varepsilon}$ is the refractive index; λ is the wavelength of the microwave radiation. If the chamber is first pumped to a high vacuum then $N_0 = 1$.

For simplicity we shall consider a plasma of uniform density with sharply defined boundaries. Eliminating the refractive index from Eq. (7.8) we obtain an expression that relates the plasma density, its linear dimension, and the phase shift, in units of 2π (that is to say, the number of interference peaks κ_i),

$$n = n_{cr} \left[1 - \left(1 - \frac{\lambda \kappa_i}{l} \right)^2 \right]. \qquad (7.9)$$

It follows from this expression that the density reaches the critical value when $\lambda \kappa_i = l$, that is to say, before the microwave beam is cut off a number of interference peaks $\kappa_i = l/\lambda$ appear on the screen of the oscilloscope. Thus, the interference pattern obtained by means of the simple interferometer can be used, in principle, to determine the plasma dimensions at the time in which the microwave signal is cut off. In discussing interferometers attention should be given to one of the most common errors that arises in the determination of the density from the number of interference peaks. The presence, between the horn and the plasma, of any channel with a refractive index appreciably greater than unity (for example, the glass wall of the chamber) can cause the appearance of additional peaks by virtue of reflection from the walls of the microwave beam reflected from the plasma; in this case the use of Eq. (7.8) is no longer valid.

The magnitude of the density and the dimensions of the plasma can be determined uniquely from interference patterns only if the variation in density is monotonic (obviously in the absence of reflections). The existence of additional peaks showing the time dependence of the density leads to the appearance of additional interference peaks which correspond to each reduction or increase in the density and which make the unique interpretation of the experimental curve essentially impossible. In order to establish a unique correspondence between the interference peaks and the density it is necessary to have additional data which indicate whether the interference peak corresponds to an increase or a reduction in density. These data can be obtained automatically in an interferometer proposed by Wharton [1, 2].

Let us consider the general features of the interference pattern that is obtained in an interferometer that has different lengths of waveguide in the reference and measurement channels and which is driven by a source with a smoothly varying frequency. In the transmission of a microwave with a fixed value of λ in the absence of plasma the phase shift is

$$\varphi_1 = 2\pi \frac{\Delta L}{\lambda}, \qquad (7.10)$$

where ΔL is the difference between the path lengths of the waves in the two channels and λ is the wavelength of the microwave signal.

If the frequency of the generator is not constant, but varies periodically, then even with a fixed dielectric constant in the medium there will be periodic changes in the phase differences. Consequently, there will be periodic variations in the microwave signal recorded by the detector. In the Wharton interferometer the phase differences causes by the change in the frequency of the microwave signal and that caused by the change in the dielectric constant in the measurement channel are both recorded. The times corresponding to interference peaks appear on the screen of the oscilloscope in the form of bright spots. The periodic sawtooth voltage pulses which control the frequency of the klystron are applied to the vertical deflection plates of the oscilloscope. The length of the sawtooth pulse is equal to the modulation period of the generator and is much smaller than the length of the horizontal sweep. With a fixed beam brightness the screen of the oscilloscope shows a sequence of sawtooth pulses.

In the Wharton interferometer the brightness of the oscilloscope beam is modulated in such a way that the luminous spot on the screen is visible only at times corresponding to interference peaks. As a result, if the difference in path length of the beams in the measurement channel and in the reference channel is held fixed the screen of the oscilloscope shows a sequence of luminous points which form horizontal dashed lines. In this case the phase shift is determined only by the magnitude of the control voltage and an interference peak of a given order appears each time the beam moves over a distance corresponding to a definite voltage. The pattern of luminous points is shown in Fig. 7.4a. The dashed curves indicate the sawtooth voltage pulses which control the oscillator frequency. The number of lines on the screen is equal to the number of interference peaks that are produced in a single excursion of the oscillator frequency. The distance between neighboring lines corresponds to a change of phase equal to 2π.

The introduction of a plasma between the detection and transmission horns causes an additional phase shift which, in turn, causes a displacement of the line by a distance

$$y = \frac{d}{\lambda} \int_{0}^{l} [N(x) - 1]\, dx, \qquad (7.11)$$

which is uniquely related to the plasma density, where d is the distance between lines.

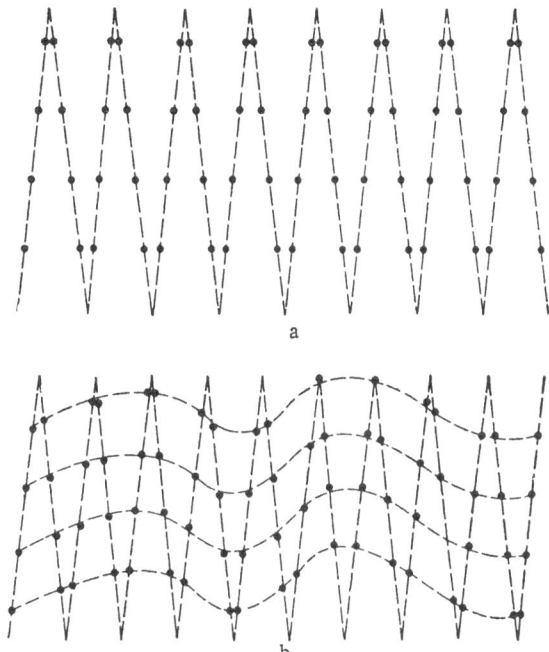

Fig. 7.4. Diagram to illustrate the operation of the Wharton
interferometer. a) Sawtooth voltage pulses used for the
sweep. The ● denotes the bright lines in the absence of
plasma. b) Interference curves produced when the plasma
density goes through the peak value twice.

A change in the plasma density in the path of the microwave
beam obviously causes a distortion of the horizontal dotted lines.
In this case the deflection of the dotted lines from the original
position at a given point corresponds to the change in the refractive index corresponding to a given instant of time. Consequently,
the shape of the dashed curve approximates a curve which describes the time dependence of the plasma density. In experiments with an inhomogeneous plasma the dashed curves denote the
change in the mean density in the path of the microwave beam.

As an example we shall consider an interferometer pattern
which is obtained with a Wharton arrangement showing the time
dependence of a density which goes through two peaks. As we have
noted above, measurements of this kind of time dependence made
with a simple interferometer cannot be used without additional in-

formation. When the plasma density is equal to zero fixed straight dashed lines appear on the oscilloscope. As the plasma density increases the dashed lines become curved and the relation between the magnitude of the displacement from the initial position and the plasma density is described by Eq. (7.11). The peak deflection corresponds to the peak density. A reduction in density causes a smaller phase shift and, consequently, a smaller deflection from the initial position. A second increase in density again increases the phase shift and the dashed curves on the screen of the os-cilloscope again indicates the increase in density (cf. Fig. 7.4b).

In contrast with a microwave interferometer that operates at a fixed wavelength the scheme being considered here for in-vestigating plasma density makes it extremely easy to establish a unique correspondence between the shift of the interference bands and the refractive index of a plasma, that is to say, the den-sity can be determined for a nonmonotonic variation of the density with time.

In working with low-density plasmas a significant phase shift can be obtained by using low-frequency waves. However, the use of long wavelengths is limited by diffraction effects which become appreciable as the wavelength approaches the dimensions of the plasma.

The minimum value of the density that can be measured by the interferometer for a fixed wavelength λ is determined by the smallest phase shift that can be measured reliably. It will be evident that this minimum value will below only for sharp in-terference peaks, that is to say, under conditions such that a small change in phase corresponds to an appreciable change in in-tensity. In a system with two interfering beams the width of the peak is large and is given by $\sim \pi$. In this respect the situation is much better in multibeam devices, for example, in a Fabry-Perot interferometer operated in the millimeter region [3, 4]. The width of the peak in a Fabry-Perot interferometer is equal to $2\frac{1-r}{\sqrt{r}}$ (r is the reflection coefficient of the mirror) and, with good mir-ros, this value can reach several hundredths of a radian. The measured phase shift is related to the density by the expression

$$n = \frac{mc^2}{e^2} \cdot \frac{\varphi}{2\lambda l}, \tag{7.12}$$

where l is the plasma dimension and φ is the phase shift.

Mirrors for a millimeter Fabry-Perot interferometer are fabricated from polystyrene which has deposited on its surface a layer of silver of thickness 200-300 Å. The transparency of the mirror is obtained by using alternate unplated band strips. The choice of the direction of the bands and the strip density are determined by the direction of polarization of the radiation and the desired transparency of the mirrors. Such a mirror can transmit a wave polarized with its electric vector perpendicular to the direction of the slits. This property of the mirror must be kept in mind in operating with a plasma in a magnetic field since the electric vector of the wave must be along the lines of force of the magnetic field.

The use of a mirror of spherical or cylindrical shape makes it possible to focus the microwave radiation in the region occupied by the plasma. This feature is of special importance when the conditions of operation do not make it possible to locate the mirror in direct proximity to the plasma. In place of two individual mirrors it is sometimes possible to use a reflecting wall. The interference pattern and method of treating the results here are the same as for a cylindrical mirror. This type of interferometer has been used widely by V. E. Golant and his coworkers [5].

The possibility of experiments on probing a plasma with microwave radiation are not limited to measurements of density. Valuable information on the state of a plasma can also be obtained by an analysis of scattered microwave radiation. In the microwave region the condition $\lambda \gg \lambda_D$ obtains so that the scattering does not occur on individual electrons, but rather on fluctuations of the density of electrons grouped around the ions. The line shape of the scattered radiation can then be used to determine the plasma temperature. Typical line shapes corresponding to different ratios of electron temperature to ion temperature are shown in Fig. 4.4. An experiment of this kind is rather difficult as a method of determining temperature.

It is somewhat easier to deal with the detection of microwave radiation scattered from a turbulent plasma in which the amplitude of the fluctuations exceeds significantly the amplitude of the thermal noise. However, in such a plasma the measurements of the line shape of the scattered radiation can only be carried out qualitatively at the present time.

The greatest interest for plasma diagnostics stems from measurements of combination sacattering [30] on plasma oscillations in which the oscillation frequency coincides with the frequency of the probing microwave. The point here is that the power of the combination scattering is linearly related to the energy density of the noise in the plasma. At first glance attention would appear to be directed toward experiments on scattering at high frequencies since the power of the scattered radiation increases as the wavelength is reduced, going as $1/\lambda^2$. Unfortunately, a theoretical analysis of the scattering coefficient at the plasma frequencies does not allow a unique determination of this quantity, that is to say, the results of the calculation are sensitive to assumptions that are made with respect to the nature of the plasma oscillations and it is almost impossible to obtain experimental verification of these assumptions. In order to eliminate contribution of the uncertainty in the experimental conditions one can use the results of measurements of the radiation power from the plasma at twice the plasma frequency. This radiation arises as a result of combination scattering of plasma oscillations off each other or, as it is usually called, the interaction between two plasmons.

Since the scattered radiation power I_s is proportional to the energy density of the plasma oscillations while the radiated power at twice the plasma frequency I_r is proportional to the square of the energy, assuming that the probability of both elementary processes is the same, we have

$$W = \frac{I_r}{I_s} \cdot \frac{I_o}{c}$$

where W is the power of the probing radiation at frequency ω_0, I_0 is the energy of the oscillation, and c is the velocity of light.

Experimental work on combination scattering has been carried on fairly widely (cf., for example, [31-33]). In [33] measurements were made of the oscillation energy of the plasma in a toroidal device used in a turbulent heating experiment. At a current of 0.5 kA typical values of the parameters were as follows: $n \sim 10^{12}$ cm^{-3} and $T_e \sim 10^2$ eV. The power of the probing radiation at a wavelength $\lambda = 3$ cm was 10 kW and the pulse length approximately 5 μsec. An analysis of the scattered radiation was carried out by means of an assembly of waveguides beyond cutoff, each of which passed microwave radiation at a wavelength shorter than a predetermined value. In order to avoid noise background due to the probing radiation a narrow band resonance filter was used. The scattered radiation and the plasma radiation were recorded by the same detector. The energy density of the plasma oscillations was found to be $\sim 0.1\ kT_e$. Similar experiments have been carried out with ion-acoustic waves [34].

The technology of microwave measurements is more or less a self-sustaining field of contemporary technical physics. Within the framework of the present section it would be impossible to consider various methods of generating and detecting microwaves. A useful acquaintance with these problems can be obtained in greater detail from a number of monographs in which the physics and engineering of microwave diagnostics is considered in detail and which contain references to original work [6-8].

§ 7.2.   Interferometer Methods

for Investigating Plasma

in the Visible Region.

Use of Lasers

In the investigation of high-density plasma, when microwave probing cannot be used, it is possible to make use of monochromatic beams in the visible and infrared regions of the spectrum. Through the use of these techniques it is possible to increase the upper limit of measurable densities by several orders of magnitude. Work with beams in this region of the spectrum is convenient in that here the refractive index is essentially independent of the direction of the magnetic field with respect to the plane of polarization of the beam. If $\omega \gg \nu_{col}$ and the electric vector of the electromagnetic wave is perpendicular to the lines of force of the external magnetic field, at high frequencies the dielectric constant is given by

$$\varepsilon = 1 - \frac{\omega_0^2}{\omega^2 \pm \omega\omega_H}, \qquad (7.13)$$

where

$$\omega_H = \frac{eH}{mc} = 1.7 \cdot 10^7\, H \text{ Oe}$$

is the electron plasma frequency. For all reasonable values of H the inequality $\omega \gg \omega_H$ is satisfied in the visible and infrared regions, that is to say, the dielectric constant of the plasma is described by Eq. (7.2) even in the presence of a magnetic field.

The use of infrared and visible light for plasma diagnostics was retarded for a long time by the absence of sources with the required power density and coherence. For this reason interferometer investigations of plasma in this spectral region were extremely rare. The situation has been changed completely, however, by the appearance of lasers, which satisfy both of these requirements: a high degree of coherence and high power within a narrow spectral range. It is sufficient to say that the monochromatic nature of the laser radiation is indicated by the factor 10^{-12}-10^{-13} while the power in a pulse is 10^8 W. In contrast with microwave beams the diameter of the beam can always be made small enough as to allow an investigation of local variations in density.

In recent years a number of papers have appeared in the literature in which the determination of plasma density has been carried out using interferometer methods based on the application of lasers, primarily gas lasers (cf. [9-12]). In spite of the fact that the radiated power of a gas laser is much smaller than that of a crystal laser (usually hundredths of a watt) operation with gas lasers is generally more convenient because of their cw operation. The generation region for gas lasers extends from the visible region up to wavelengths of tens of microns.

One of the simplest interferometer arrangements that makes use of a laser is shown in Fig. 7.5. The beam from the laser is directed along the axis of a Fabry-Perot interferometer, between the plates of which is the object being investigated, the plasma. The radiation transmitted through the Fabry-Perot interferometer is recorded by a photomultiplier which is connected to an oscilloscope [7]. The dependence of the transmission at the center of the interferometer on the dielectric constant of the medium exhibits sharp peaks, as is well known.

The change in the plasma density corresponding to a phase shift 2π, that is to say, the distance between neighboring peaks, is given by the expression

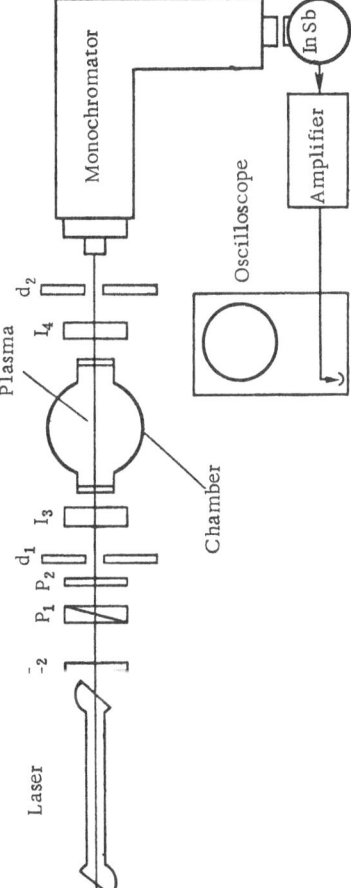

Fig. 7.5. Diagram of an interferometer that makes use of a gas laser: I_1 and I_2 are the laser mirrors; I_3 and I_4 are the mirrors of the Fabry-Perot interferometer; P_1 is the polarizer; P_2 is the quarter-wave plate; d_1 and d_2 are diaphragms.

$$\Delta n = \frac{\pi m c^2}{e^2 l \lambda}, \quad \text{or} \quad \Delta n = \frac{1.1 \cdot 10^{13}}{\lambda l}. \tag{7.14}$$

Here l is the path length of the beam in the plasma and λ is the wavelength in centimeters.

Most of the work on interferometer determination of plasma density has been carried out with helium–neon lasers. This laser produces bright lines both in the visible region of the spectrum ($\lambda = 0.63 \mu$) and the infrared region ($\lambda = 3.39 \mu$). Since the method of determining the plasma density from the measurement of the number of interference peaks has a higher sensitivity, the greater the wavelength of the radiation, in investigations of modestly dense plasmas most frequently the line $\lambda = 3.39 \mu$ is used. Another advantage of the long wavelength is the less stringent requirement on the adjustment of the mirrors and the smaller effect of mechanical vibrations, which become especially important in the practical use of interferometers. The most convenient detector for radiation in this spectral region is a photoresistive InSb element operated at liquid-nitrogen temperature.

If the laser radiates at two wavelengths, $\lambda = 0.63 \mu$ and $\lambda = 3.39 \mu$, simultaneously, it is not necessary to use an infrared detector in detecting the interference peaks for the 3.39 μ line since it is possible to use the same photomultiplier which records the intensity of the red line at 6.3 μ. In this case, the interferometer arrangement shown in Fig. 7.5 can be simplified considerably. The mirror I_3 is removed and the mirror I_2 serves simultaneously as the mirror of the laser and as the mirror in the interferometer. Furthermore, the detector, i.e., the photomultiplier, is located outside the semitransparent spherical mirror I_1. Owing to the fact that both the red and the infrared lines arise from transitions from the same excited level, weak modulation of the intensity of the infrared line causes a modulation of the red line, the intensity of which is detected by the photomultiplier. In order to resolve interference fringes corresponding only to the infrared line, it is necessary to insert between the mirrors I_2 and I_1 a filter which is transparent for the infrared region of the spectrum, but opaque for the visible region. A thin germanium plate can be used. Interference of the red light without side effects at the wavelength 3.39 μ can be obtained if a glass plate is used as a filter. Thus, in this experimental configuration it is possible to have independent monitoring of the measurements using lines in different regions of the spectrum.

The sensitivity of the interferometer is increased if spherical mirrors are used instead of plane mirrors in the Fabry-Perot interferometer. In the cavity formed by such mirrors, in addition to the longitudinal modes, there are also transverse modes of oscillation. The expression that relates the change in the density with the number of interference fringes for this case is of the form

$$\Delta n = \frac{1.1 \cdot 10^{13}}{\lambda l} \left[\Delta q + \frac{\Delta m + \Delta p}{\pi} \cos^{-1} \left(1 - \frac{l}{r} \right)^{1/2} \right], \qquad (7.15)$$

where q, m, and p are the orders of the interference peaks corresponding to various modes. As the radius of curvature of the mirror approaches infinity the intensity of modes corresponding to the transverse modes tends to zero and Eq. (7.15) goes over to Eq. (7.14) for the straight Fabry-Perot interferometer, as expected.

In [13] the radius of the mirror is taken to be 1 m while the plasma dimension $l = 50$ cm. The indicated values of r and l correspond to $\cos^{-1}[1 - (l/r)]^{1/2} = \pi/4$. Consequently, under the conditions of this experiment each interference fringe corresponds to a change in the density which is four times smaller than in an interferometer with plane mirrors. However an interferometer with spherical mirrors has the important disadvantage that the results of the measurements permit a unique analysis only in experiments with a sufficiently uniform plasma of known geometry.

Obviously the minimum density that can be measured in an interferometer with plane mirrors corresponds to the minimum displacement of the interference rings which can be detected photographically or with a photomultiplier. The value of density is very sensitive to the shape of the interference curve. In an interferometer with a low-quality factor, in which the interference curve is approximately sinusoidal and the modulation is small, $\frac{I_{max} - I_{min}}{I_{max}} \ll 1$, it is extremely difficult to measure a phase shift smaller than a tenth of 2π. The minimum measurable density is appreciably smaller for interferometers in which the mirrors have high coefficients of reflection and the loss of light due to the absorption of penetration through the window of the vacuum chamber is reduced to a minimum. In this case the interferometer curve consists of sharp transmission peaks and choosing the operating point on the sharpest slope of the interference peak means that it is possible to measure the density $(2-3) \cdot 10^{12}$ cm^{-3} [14, 29]. In order to correct the distance between the mirrors so that the adjustment of the device corresponds to the chosen operating point (with no plasma), it is found to be most convenient to use a ring of barium titanate, the height of which is controlled by means of an electric field.

In choosing the actual form of the interferometer one should keep in mind the fact that a multibeam interferometer device does not always provide the required time resolution. In a Fabry-Perot interferometer the time constant is determined essentially by the time which is required to fill the device with light. The time constant is

$$\tau = \frac{\pi r}{1 - \sqrt{r}} \cdot \frac{l}{c}, \qquad (7.16)$$

where r is the reflection coefficient of the mirror and c is the velocity of light.

Modulation of the light reflected from the interferometer in the laser cavity is equivalent to periodic variation of the reflection coefficient of the laser mirror which, in turn, causes a periodic variation of the intensity of the laser output. The time for retuning a gas laser, as observed experimentally, is of the order of 1 μsec. Thus, the time resolution is determined by the coupled system, consisting of the generator and the interferometer.

In devices with high time resolution, the negative feedback is usually reduced by using a polarizer and a mica slab [thickness equal to a quarter wavelength of the radiation $\lambda/4(N_1 - N_2)$] between the laser and the interferometer. Here N_1 and N_2 are the refractive indices of mica for light polarized along the principal axes of the crystal [15]. After a double pass through the slab (before striking the mirror of the interferometer and after reflection from the mirror) the plane of polarization of the polarized beam is rotated through an angle of 90° and the beam cannot pass through the polarizer a second time; consequently, it does not enter the laser cavity. This system almost completely avoids the feedback between the interferometer and laser and thus eliminates the effect of the oscillator time constant on measurements. In working with a plasma whose parameters vary much more rapidly than the time constant of the laser [but slower than the resolution time of the interferometer (7.16)], there is no need to use the quarter-wave plate. This is the case because the laser mode of operation cannot change during the time of the measurement of the density.

Small densities can be measured by a method in which two lasers are used simultaneously. The resonator of one of the lasers contains the plasma that is being studied. The reduction of the refractive index caused by the plasma leads to a change in the optical length of the resonator, and, consequently, to a shift in the frequency of this laser. The second laser is used as a fixed-frequency source. By beating the two lasers and detecting the beat with a photomultiplier and a spectrum analyzer, it is possible to measure the frequency shift. This method can be used to measure densities down to 10^9 cm^{-3}.

The determination of the electron density by interferometer methods is sometimes complicated by the phase shift due to the presence of a large amount of neutral gas whose density varies with time [16]. When a microwave-diagnostic system is used the contribution of the neutral gas to the phase shift is neg-ligibly small. On the other hand, when using electromagnetic radiation in the visible region one should always keep in mind the dependence of the refractive index on the density of neutral atoms. The refractive index of a nonionized medium is

$$N = 1 + \frac{2\pi e^2 n_0}{m} \sum \frac{f_k}{\omega_k^2 - \omega^2}, \tag{7.17}$$

where n_0 is the density of molecules; ω_k is the frequency of the radiation; f_k is the oscillator strength for the appropriate transition.

In order to determine the refractive index of a nonionized gas in the visible region we can use the expression

$$N = 1 + \alpha \left(1 + \frac{\beta}{\lambda^2}\right) n_0. \tag{7.18}$$

Here λ is the wavelength in angstroms. The coefficients α and β are given in Table 7.1 for various gases.

TABLE 7.1. Refractive Indices

Gas	$\alpha, \times 10^{-24}$	$\beta, \times 10^{5}$	Gas	$\alpha, \times 10^{-24}$	$\beta, \times 10^{5}$
H	5.0	7.5	Ar	10	5.6
He	1.3	2.3	Kr	16	7
O_2	10	5.1	Xe	25	10
N_2	11	7.7	Hg	33	22.7
Ne	2.5	2.4			

By exploiting the wavelength dependence of the refractive indices of the neutral gas and the ionized gas it is possible to determine separately the density of electrons and molecules in a plasma by carrying out measurements at two wavelengths, for example, in the visible and in the infrared. Comparing Eq. (7.17) with the formula for the refractive index of an ionized gas,

$$N = 1 - 9 \cdot 10^{-30} \lambda^2 n_e \qquad (7.19)$$

(λ is given in angstroms), one is easily convinced that in the visible region (with equal densities of neutral and charged particles) the change in the refractive index due to the charged particles is approximately an order of magnitude larger. Consequently the effect of the neutral gas must be taken into account when its density becomes comparable with the electron density in the plasma.

Interferometer methods have been widely used for studying plasmas in the visible region of the spectrum since the appearance of powerful lasers, although the successful application of interference devices for the determination of densities has been known for more than ten years [17, 18]. Apparently the first use of the interference of light in the determination of plasma density is given in the work of G. G. Dolgov-Savel'ev and S. L. Mandel'-shtam [17] on the emission from a high-pressure spark. The spark being investigated was located within a Rozhdestvenskii interferometer, in which interference bands are used to determine the spectral dispersion of the light without plasma. The source of light is a spark with electrodes fabricated from magnesium. A diagram of the experiments is shown in Fig. 7.6. The spark source I_1 is located at the focus of the objective O_1 which, through an interference filter, directs the light beam into the interferometer formed by the four semitransparent mirrors. In one arm of the interferometer, in the plane of localization of the inter-

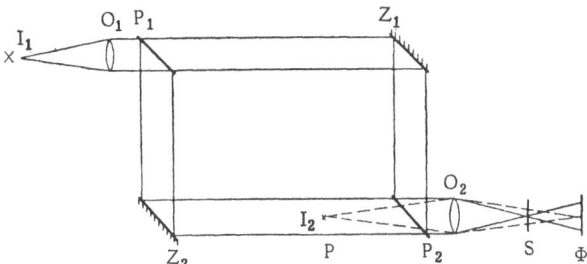

Fig. 7.6. Diagram of the Rozhdestvenskii interferometer for the determination of plasma density: P_1 and P_2 are semi-transparent plates; Z_1 and Z_2 are mirrors; O_1 and O_2 are objectives; Φ is the photographic plate; S is the slit; I_1 is the illuminating spark; I_2 is the image of the spark. The solid lines show the ray paths in the interferometer and the dashed lines show the ray paths in the plasma being investigated.

ference bands P, is located the spark being investigated I_2. The objective O_2 focuses the plane P on the film. The illumination of the spark being investigated makes a small contribution to the exposure of the film by virtue of the slit S in the principal focus of the objective O_2. The width of the slit is equal to the dimension of the image of the spark I_2 on the film. The slit makes it possible to obtain a bright image of the interference bands due to the plasma when the bright spark I_2 is fired.

Assume that the plasma is not cylindrical (as is the case of a spark) but is plane. Also, assume that the density gradient is nonzero only along an axis parallel to the interference band which is observed in the absence of the plasma (x axis). In this case the shape of the interference bands in the presence of the plasma reflects the spatial density distribution. Under these conditions the displacement of the interference band by a distance ΔZ corresponds to a change in the electron density (at a given point on the axis) by an amount

$$\Delta Z = \Delta Z_0 \frac{e^2}{2\pi m c^2} \lambda n l. \tag{7.20}$$

Here ΔZ_0 is the distance between interference bands without the plasma; λ is the wavelength; l is the length of the beam path in the plasma.

In the case of cylindrical symmetry, in which the plasma density depends only on the distance from the axis of symmetry, the curve showing the spatial density distribution can be found from the following simple operation; the region occupied by the plasma is divided into circular zones of equal thickness and the detected displacement of the interference bands is treated as the total effect of the change in density of those circular zones which intersect the interference beam. If the true density distribution can be approximated by a series of step functions such that the density can be regarded as being constant within each of the circular zones, the measured value $\int ndx$ can be replaced by a summation of a finite number of terms. For example, for the beam denoted by the dashed line in Fig. 7.7, the quantity $\int_l ndx$ is approximated as follows:

$$\int_l ndx = 2\Delta l_1 n_1 + 2\Delta l_2 n_2 + 2\Delta l_3 n_3 + \Delta l_4 n_4, \qquad (7.21)$$

where n_j is the mean density within a given circular zone and Δl_j is the path traversed by the beam within the zone.

Dividing the region occupied by plasma into j circular zones and determining the quantity $\int ndx$ j times so that points for which the displacement of the interference bands differ from each other by a distance corresponding to the width of circular zone, we can form j equations of the form of (7.21) with the unknowns n_1, n_2, n_3,..., n_j. The solution of this system of linear equations can be used to determine the mean value of the density within each of the circular zones and thus, to find the required radial distribution of density in the axially symmetric plasma column.

In recent years a number of papers have appeared in which interference methods based on the use of pulsed flash lamps have been used to investigate density distributions in moving plasmoid and to study shock waves and the flow of a plasma around solid obstacles [19, 20]. The change in the density in flow of a plasma around a solid obstacle is determined by the curvature of the interference bands; these can be observed at a given instant of time with a pulsed light source by means of a rotating mirror.

The ruby laser is found to have significant advantages as compared with other pulsed light sources in interference experiments

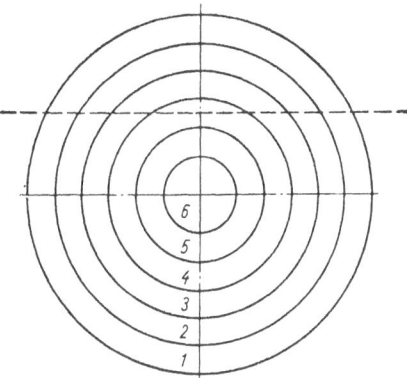

Fig. 7.7. Extrapolation of the true radial
distribution of the plasma in steps.

on the determination of plasma density in a plasma with rapidly
varying parameters. In this device the exposure time (10-20
nsec) is appreciably smaller than that of other light sources.
The short pulse length and the high intensity of the flash of monochromatic light make it convenient to obtain photographs of the
interferograms corresponding to any given instant of time. A
Fabry-Perot interferometer has been used successfully in conjunction with a ruby laser by a number of workers to determine the
radial density distribution in the Scylla device [21-23]. The interferograms obtained by photographing the apparatus from the
end faces provide a detailed method of investigating the compression process of a plasma by an external field and to examine the
various instabilities that develop during compression.

New possibilities in the technology of interference photography have appeared with the development of holography as applied to interferometers [24, 25]. In this method we do not require exact adjustment or precision lenses. Instead, the rigid
requirements are now imposed on the radiation source itself. The
radiation source must exhibit a high spatial coherence, must be
highly monochromatic, and must be characterized by a high intensity. All of these requirements are easily satisfied by a ruby
laser operated in the pulsed mode.

We now consider briefly the physical principles of holography,
using the example of a two-beam system. The radiation from the

laser is divided by means of a semi-transparent mirror into two beams, which are incident on a photographic film at different angles. One of these (Fig. 7.8a) is directed at an angle of 90° with respect to the film and is the reference beam. In the path of the second beam is located the object being investigated (for example, a plasma); the refractive index of this object does not differ greatly from the refractive index of air. After exposure of the film and development one obtains a system of bands corresponding to equal differences in path length; these form a kind of diffraction pattern. Because of the rapid variation in path between one point of the film and another the distance between interference bands is found to be rather small, amounting to 10^{-3} cm in the typical case.

If we now illuminate the hologram obtained in this way the reproduced image is formed by waves of the first-order diffraction pattern (cf. Fig. 7.8b). It is interesting to note that the first two diffraction orders are not identical. One of the diffraction orders consists of waves which appear to have originated from an object located at the object position of the hologram. These waves produce an image which is very similar to the virtual image formed in a mirror. The other first-order wave pattern is an exact copy of the original wave but with inverted curvature. These give a real image which can be directly photographed without a lens by placing a photographic film at the point at which the image is located. The zero-order diffraction wave is propagated in the same direction as the light incident on the hologram and can be regarded as the attenuated part of the incident wave.

In order to obtain an interference pattern which corresponds to the difference in the ray length in the object being investigated, the hologram is produced by taking a double exposure on the same film: once in the presence of the object (plasma) and the second time without it. The resulting hologram represents a superposition of holograms taken with and without the object and contains information on the additional difference in path length that arises because the beam passes through the object. At points at which the additional difference in path length comprises a whole number of wavelengths the diffraction pattern exhibits greater contrast. At points where the difference in wavelength is equal to a half-integral number of path wavelengths there is a continuous blackening since the interference bands of the second exposure lie precisely between the bands of the first.

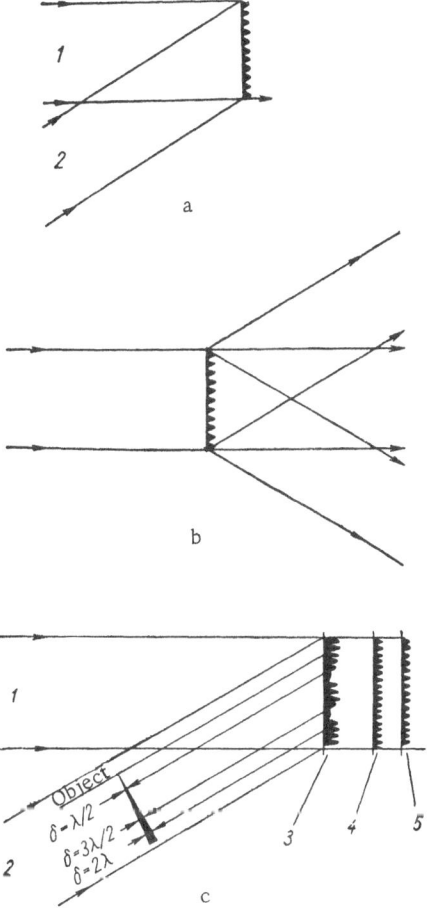

Fig. 7.8. Hologram method for obtaining an
interferogram. a) Scheme for obtaining a
hologram; b) scheme for reproduction of the
image; c) formation of the hologram of the
interference image: 1) reference ray; 2)
probing ray; 3) hologram obtained after two
exposures; 4) hologram obtained by exposure
without objective; 5) hologram exposed in the
presence of the object.

Diffraction scattering of light does not occur at these places and these points correspond to the black bands on the reproduced image. A diagram showing the formation of the hologram of a wedge in a double exposure is shown in Fig. 7.8c.

The interference bands that arise because of the additional difference in ray length in passage through the object are not visible directly on the hologram since the maxima corresponding to each of the two exposures are located so close to each other that the hologram appears to be uniformly black. Another pattern is observed in the first diffraction order. Here, each interference maximum corresponds to a whole series of these maxima whose contrast is increased in the second exposure. On the other hand, the interference minima correspond to those portions of the hologram in which the maxima of the second exposure lie between the maxima of the first exposure. As a result, the contrast of the maxima in these sections is much weaker and these lose the properties of a diffraction grating. This interferogram of the object has sufficient contrast if the position of the nodes of the apparatus can be fixed reliably and if the production of the hologram and the reproduction of the image are carried out in the same geometry.

Interference holograms were first used for the investigation of plasmas by A. N. Zeidel and his colleagues [26-28] in experiments devoted to an investigation of a laser-produced plasma with a density $\sim 10^{19}$ cm^{-3}. Exposing the film in a two-beam interferometer twice (once in the presence of the plasma of the laser spark and the second time without it) the authors were able to obtain holograms which, after development, made it possible to fix the density front in the form of interference bands of equal difference in path length. Before one of the two exposures a thin wedge was placed in the path of the reference beam which, in the absence of the plasma, produced linear interference bands. The presence of plasma produces a distortion of the bands in accordance with the additional difference in path length. In this case, in the investigation of the distribution of density in absolute units the difference in path length introduced by the spark is computed from the displacement (more precisely the curvature) of these bands.

In these same experiments it was possible to obtain an interferogram corresponding to a time interval of $4 \cdot 10^{-8}$ sec using the delay of light reflected by a system of semitransparent and opaque mirrors. Light beams that traverse long paths because of reflections from special mirrors then intersect the plasma with different delays. The resulting interference images then correspond to different delay times.

Schlieren photography is used to obtain data on the shape and localization of the plasma regions that exhibit a high gradient in the index of refraction, that is to say, regions with a high density gradient. This term comes from the word "schliere," which is frequently used in connection with the inhomogeneity of optical glass, which produces bands of an optical image. The principle of schlieren photography is based on the deflection of a light beam as it passes through an optically inhomogeneous medium. The angle of deflection of a beam in a medium that exhibits a gradient in the refractive index is given by

$$\varphi = \int_0^l \frac{1}{N} \cdot \frac{\partial N}{\partial x} \, dz, \qquad (7.22)$$

where N is the refractive index and l is the path length of the beam in the plasma; the original direction of the beam coincides with the z axis, the x axis is in the direction of the density gradient, and the angle φ lies in the xz plane.

Let us consider an optical system in which a parallel beam of light first passes through the object being investigated and then strikes the lense O_1 and, passing through the focus, again diverges (Fig. 7.9). In the path of the diverging beam there is a second lense O_2 whose focus is the same as the focus of lense O_1. If, in passing through the object, the trajectory of the rays is not varied, the lense O_2 will again form a parallel beam which illuminates the screen uniformly. This is essentially the pattern that appears when a transparent object is located in the path of the beam and the gradient of the refractive index is nonvanishing. In order to obtain a schlieren photograph it is necessary to take account of the beam which changes its direction of propagation because it propagates in a region with a strong gradient. For this purpose, at a small distance from the focus there is located an opaque plate with a sharp rectilinear edge (wedge) which cuts off the beam that deviates in a given direction downward (Fig. 7.9). To be definite we shall assume that the refractive index in some region in-

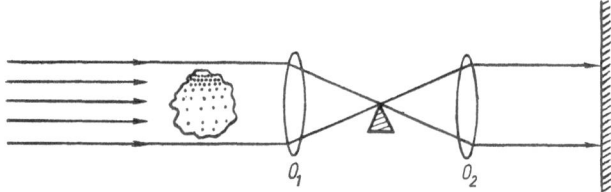

Fig. 7.9. Diagram showing schlieren photography of a plasma.

creases in the vertical direction (going upward); then the beam
passing through this region is deflected downward and will pass
below the focus of the lens. By locating the wedge below the focus
so that its sharp edge is perpendicular to the plane of the image,
we can cut off all rays that are deflected by an angle greater than
some critical value. Then those points of the image of the screen
which correspond to rather high values of the gradient in the refractive index and in which the refractive index increases going
vertically upward will be found to be darker.

The application of schlieren photography to the diagnostics of
plasmas was retarded for a long time by the lack of powerful sources
of radiation which could exceed the brightness of the plasma. With
the advent of lasers this problem no longer exists. Through the
use of lasers it is possible to produce a short flash which can be
used to take a reasonable picture with an exposure of the order of
nanoseconds.

The minimum value of the gradient in the density that can be
observed by schlieren photography is limited by diffraction effects
at the wedge. This minimum value can be estimated roughly by
determining the increment in the displacement of the ray in the region of the focus at the wavelength in question,

$$\nabla n_e \approx \frac{2 \cdot 10^{13}}{lD\lambda},\qquad(7.23)$$

where l is the path length of the beam in the plasma; D is the
diameter of the lens; λ is the wavelength in centimeters.

The localization of regions with sharp inhomogeneities in a
plasma are clearly seen on photographs obtained by the shadow

method. If the object to be investigated is located in the path of the beam of parallel rays which illuminates the screen (or the film) a shadow photograph of the object is obtained on the screen. Since the optical length of the plasma is rather large for nonresonance radiation, the shadow image of the plasma will not be due to absorption effects, but rather to the change in the trajectory of the beam due to the inhomogeneity in the refractive index. In contrast with schlieren photography, the shadow image does not allow us to obtain an image of the regions with constant gradient since the presence of the gradient in the density of the plasma shifts the entire image as a whole. If, however, the gradient is not constant, that is to say, if the derivative $\partial^2 n / \partial r^2$ is nonvanishing, the illumination of the screen becomes nonuniform. In this case the weakening of the illumination is proportional to the derivative of the density gradient,

$$dJ = -\,4.5 \cdot 10^{-14}\, JL\lambda^2 \int_0^l \left(\frac{\partial^2 n_e}{\partial x^2} + \frac{\partial^2 n_e}{\partial y^2} \right) dz,$$

where L is the distance between the object and the film and λ is the wavelength in centimeters.

Regions with varying gradient can be regarded as a particular kind of optical system which focuses or defocuses a beam. For example, a region with a negative second density derivative is analogous to a focusing lens.

In relief shadow photography, special importance attaches to the choice of the source. In the ideal case the source should produce a beam of nonintersecting rays in order to eliminate the possibility of half shadows. This requirement is satisfied by a laser operated in the pulsed mode.

References

1. C. B. Wharton et al., in: Proc. 2nd International Conference on the Peaceful Uses of Atomic Energy, Geneva, 1958.
2. E. P. Gorbunov, in: Plasma Diagnostics [in Russian], Gosatomizdat, Moscow, 1963, p. 68.
3. R. J. Primich and R. A. Hayami, Trans. IEEE, MTT-12, 33 (1964).
4. V. Ya. Balakhanov, V. D. Rusanov, and A. R. Striganov, Zh. Tekh. Fiz., 35:127 (1965) [Sov. Phys. — Tech. Phys., 10(1):96 (1965)].

5. A. I. Anisimov, et al., Zh. Tekh. Fiz., 35:2042 (1965) [Sov. Phys. — Tech. Phys., 10(11):1566 (1966)].

6. U. D. Raizer and I. S. Shpigel', Usp. Fiz. Nauk, 64:641 (1958).

7. V. D. Rusanov, Modern Methods of Plasma Research [in Russian], Gosatomizdat, Moscow, 1962.

8. A. V. Chernetskii, O. A. Zinov'ev, and O. V. Kozlov, Instruments and Methods in Plasma Research [in Russian], Atomizdat, Moscow, 1965.

9. D. E. Ashby and D. F. Jephcott, Appl. Phys. Lett., 3:13 (1963).

10. U. Ascoli-Bartoli, et al., 2nd International Conference on Plasma Physics and Controlled Thermonuclear Fusion Research, Culhan, England, 1965; IAEA, Vienna, 1966.

11. H. Malamud, Rev. Sci. Instrum., 36:1507 (1965); 34:121 (1963).

12. E. B. Hooper and G. Bekefi, Appl. Phys. Lett., 7:133 (1965).

13. J. B. Gerardo and J. T. Verdeyen, Appl. Phys. Lett., 3:121 (1963).

14. V. V. Korobkin, in: Plasma Diagnostics [in Russian], Gosatomizdat, Moscow, 1963, p. 36.

15. J. B. Gerardo, J. T. Verdeyen, and M. A. Gusinow, J. Appl. Phys., 5:2146 (1965).

16. E. A. McLean and S. A. Ramsden, Phys. Rev., 140:1122 (1965).

17. G. G. Dolgov-Savel'ev and S. L. Mandel'shtam, Zh. Eksp. Teor. Fiz., 24:691 (1953).

18. U. Ascoli-Bartoli, et al., Nuovo Cimento, 18:1116 (1960).

19. É. P. Kruglyakov, V. K. Malinovskii, and Yu. E. Nesterikhin, Magnitnaya Gidrodinamika (Magnetohydrodynamics), 2:31 (1965).

20. S. A. Ramsden and E. A. McLean, Nature, 194:761 (1962).

21. E. Fünfner, et al., Z. Naturforsch., 17a:967 (1962).

22. F. C. Jahoda, et al., Los Alamos Sci. Lab. Rep. LAMS-3004 (1963).

23. V. E. Mitsuk, V. I. Savos'kin, and V. A. Chernikov, Zh. Tekh. Fiz., 35:1156 (1965) [Sov. Phys. — Tech. Phys., 10(6):891 (1965)].

24. É. Leit and Yu. Upatnieks, Usp. Fiz. Nauk, 87:521 (1965).

25. L. O. Heflinger, R. F. Wnuerker, and R. E. E. Brooks, Appl. Phys., 37:642 (1966).

26. G. V. Ostrovskaya and Yu. I. Ostrovskii, ZhETF Pis. Red., 4:121 (1966) [JETP Lett., 4(4):121 (1966)].

27. A. N. Zaidel', et al., Phys. Lett., 23:81 (1966).

28. Yu. V. Ashcheulov, et al., Phys. Lett., 24A:61 (1967).

29. G. G. Managadze, I. M. Podgornyi, and V. D. Rusanov, Report at the International Symposium on Shock Waves, Novosibirsk, 1967.

30. A. I. Ivanov and D. D. Ryutov, Zh. Eksp. Teor. Fiz., 48:1366 (1965) [Sov. Phys. — JETP, 21(5):913 (1965)].

31. V. E. Golant, Microwave Diagnostics of Plasmas, Nauka, Moscow, 1968.

32. I. I. Chen, R. F. Leheny, and T. C. Marshall, Phys. Rev. Lett., 15:184 (1965).

33. B. A. Demidov and C. D. Fanchenko, Atomnaya Énergiya, 20:516 (1966) [Atomic Energy, 20(6):597 (1966)].

34. V. Arunasalam and S. C. Brown, Phys. Rev., 140A:471 (1965).

Chapter 8

Particle Methods for Plasma Diagnostics

§ 8.1. Use of a Beam of Charged Particles to Measure Electric Fields in a Plasma

A basic advantage of the method considered here, as compared with electric probes, is the fact that an ion beam of low intensity creates a very small perturbation of the plasma. A beam of charged particles that passes through a plasma can be used to investigate both constant and time-varying electric fields. The frequency and amplitude of high-frequency oscillations can be determined directly from an analysis of the energy distribution or the spatial localization of a beam that is transmitted through the plasma at different instants of time. The method by which the experiment is carried out depends very much on the nature of the problem, on the configuration of the magnetic field, and on other experimental conditions, so that it is extremely difficult to formulate general principles for the use of a beam of charged particles. We shall limit ourselves to the analysis of several particular cases. We first consider experiments on the determination of the azimuthal and radial components of the electric field that arises in an axially symmetric magnetic-mirror device when it is filled with plasma. The beam of electrons moves along the lines of force of the magnetic field in the absence of plasma; in the presence of an electric field these electrons experience a drift in the crossed fields. By locating an electron gun

in the region of one of the mirrors and a fluorescent screen at
the other it is possible to obtain visual or photographic indica-
tions of the displacement of the position of the beam due to the
components of the electric field which are perpendicular to the
lines of force of the magnetic field. The system essentially comprises an oscilloscope with the role of the deflection plates being
played by the stationary and high-frequency electric fields that
arise in the plasma. In the absence of the plasma the beam of
electrons moves along the lines of force of the magnetic field and
produces a luminous dot on the screen. The screen is protected
by a thin aluminum foil from the incident light and is also furnished with a system of grids, which prevent slow charged particles from striking the screen.

The presence of drift motion due to a stationary electric field
leads to a displacement of the luminous spot on the screen by a
distance given by

$$\Delta x = \frac{cE_y}{H_z} \cdot \frac{l}{v}, \tag{8.1}$$

where l is the path length of the beam in the plasma; E_y is the component of the electric field perpendicular to the lines of force; v is
the velocity of the electrons along the lines of force, which is
determined by the accelerating potential of the electron gun and
by the potential of the given magnetic tube with respect to the walls
of the chamber. The possibility of determining the potential of the
magnetic tube by means of a beam of charged particles will be considered below.

If the electron velocity is known, by fixing the position of the
beam from the spot on the screen in the presence of plasma and
without it is possible to determine both the radial and azimuthal
components of electric field. In accordance with Eq. (8.1), the displacement of the beam in the radial direction determines the magnitude and direction of the azimuthal component E_φ while the displacement in angle can be used to obtain E_r.

By observing the drift of electrons in crossed fields it is possible to determine the high-frequency component of the electric
field as well as the constant component. In a high-frequency field
the spot on the screen describes a curve which corresponds to the

time variation of the magnitude and direction of the electric field. If there are regular oscillations a stable closed curve is observed on the screen. The limit of the time resolution of the method is determined by the transit time of the electrons.

In the Ogra device the radial electric field close to the walls was measured by a different method, based on the use of a beam of charged particles [1]. On the basis of a solution of the equations of motion for a singly charged cesium ion in cross-uniform electric and magnetic fields it is possible to compute the ion energy that is required for it to reach a detector located at some fixed distance from the source. The source of cesium ions provides a beam directed perpendicularly to the wall and, consequently, perpendicular to the lines of force of the magnetic field. The source and detector are located at the side walls of the chamber. By fixing the voltage on the accelerating electrode of the source and using the results of the calculation it is possible to determine the radial component of the electric field. This method is rather limited in application since it can only be used when the spatial distribution of the field is known; for this reason it has not been widely used in plasma diagnostics.

The determination of the plasma potential with respect to the walls of a chamber can be determined by a method based on the retardation of a beam of electrons in which the electron energy becomes smaller than eV. The further development of the method in [2, 3] has made it possible to determine the potential of a plasma from the transit time of a charged particle. In operation with a plasma in a magnetic field the velocity of the beam must be parallel to the lines of force. The transit time is given by the expression:

$$\tau = \sqrt{\frac{M_i}{2Z_l}} \int_0^l \frac{dx}{\sqrt{V_0 + V(x)}} , \qquad (8.2)$$

where Z_l is the ion charge; M_i is the ion mass; l is the plasma dimension; V_0 is the accelerating potential in the ion source; $V(x)$ is the potential distribution in the plasma along the path of the beam. If the potential drop is essentially completely concentrated within a Debye radius, which is much smaller than l, (8.2) as-

assumes the simple form

$$\tau = \frac{l}{v_i},$$

where

$$v_i = \sqrt{\frac{Z_i(V_0 + V_{\mathrm{pl}})}{M_i}}.$$

In [3] a modulated beam was directed through a plasma along the lines of force of a magnetic field. The transit time was determined from the phase shift of the modulation signal in the current at the collector. The system was calibrated by passing the beam through a potential well formed by a series of grids.

In setting up an experiment of this kind one must keep in view the fact that the transmission of a beam through a plasma can result in the retardation of the beam as a result of the interaction of the beam with collective plasma oscillations that are excited by the beam itself (cf., for example, [4]). The growth time for the two-stream instability (in the linear approximation) can be estimated from

$$\tau_{\mathrm{nu}} = \sqrt[3]{\frac{n}{n'}}\frac{1}{\omega_0}, \tag{8.3}$$

where n is the plasma density, n' is the electron density in the beam, and ω_0 is the plasma frequency. In order to avoid effects due to the two-stream instability the situation must be such that τ_{nu} is much larger than the electron transit time.

§ 8.2.   Measurement of the Density

of Neutral Particles and Charged Particles

The transmission of a beam of atoms or ions through a plasma results in atomic and electronic collisions which cause a change in the charge state of the beam particles. For a beam with specified parameters the change in the charge state is a function of electron temperature and density as well as the density of neutral gas. By choosing the beam parameters appropriately it is possible to set up conditions in which the probing of the plasma by the beam can be used to establish uniquely (or almost uniquely) the relation between the charge state and one of these quantities [5-7].

We consider the change in the charge state of a monoenergetic atomic beam, consisting of neutral atoms of hydrogen and protons, when it is transmitted through a hydrogen plasma which contains a significant number of neutral atoms. In this analysis we will neglect Coulomb scattering and ionization losses of the particles in the transmission of the beam through the plasma. The first simplification is justified because of the small value of the mean square of the scattering angle over a wide range of plasma density. This quantity is given by

$$\overline{\theta} = 2 \left(\frac{Z_1 Z_2 e^2}{Mv^2} \right) \sqrt{2 \pi n L l}, \tag{8.4}$$

where l is the path length of beam in the plasma with density n; L is the Coulomb logarithm (L $\sim$ 15); M is the mass of the particles in the beam. For protons with an energy of 10^4 eV transmitted through a plasma characterized by $nl = 3 \cdot 10^{16}$ the mean square scattering angle is 1.5°. As a rule, this angle holds under actual experimental geometries. The effect of scattering and possible control experiments to evaluate scattering have been investigated in detail [8]. The validity of the second simplification is due to the small effective cross sections for ionization and excitation.

We shall make two further assumptions which, while not really limiting the generality of the problem, will facilitate the analysis considerably: 1) we assume that the velocities of the particles in the beam are much greater than the thermal velocities of the ions in the plasma, and 2) all calculations are carried out for an ion energy $W_0 = 10^4$ eV, in which case the cross sections for charge exchange on free atoms and atoms bound in molecules are insignificant.

We now list the basic elementry processes which lead to a change in the charge state of an atomic beam transmitted through a plasma. The cross sections for these elementry processes have been investigated in detail by a number of authors (cf., for example, [9-13]).

1. Capture by a fast proton of an electron in the interaction with neutral atoms in the plasma. The probability of this process

in the transmission of an ion over a path length dl in a medium with a density of neutral atoms n_0 is $n_0 \sigma_{10} dl$. The cross section σ_{10} corresponding to an energy $W = 10^4$ eV will be taken as 10^{-15} cm.

2. Stripping of an electron associated with a fast atom in the interaction of the atom with neutral atoms in the medium. The probability of the process is $n_0 \sigma_{01} d l$. The cross section σ_{01} for $W_0 = 10^4$ eV will be taken as $6 \cdot 10^{17}$ cm^{-2}.

3. Stripping of an electron associated with a fast atom on plasma protons. The probability of the process is $n_i \sigma_{10} dl$. The effective cross section σ_{10} is the same as the cross section for the process in (1).

4. Ionization of fast atoms by plasma electrons. The probability for this process is $n_e \dfrac{<\sigma_i v_e>}{v_0} \ dl.$

The maximum value of the cross section for ionization by electrons σ_i is reached at an energy of 100 eV and can be $7 \cdot 10^{17}$ cm^{-2}. Here, and in the further development of the method of probing a plasma by atomic beams, we shall make use of the following notation: n_i, n_e, and n_0 are the densities of the ions, electrons, and neutral atoms in the hydrogen plasma, respectively; N_i and N_0 are the concentration of ions and atoms in the probing hydrogen beam; v_0 is the velocity of the beam particles; l is the path length of the beam in the plasma. Comparing the contribution of the processes (3) and (4) and assuming $n_i = n_e$ on the basis of plasma neutrality, we see immediately that the neglect of ionization of the beam by plasma electrons leads to an error in the determination of the density which is less than ten per cent. Thus, the determination of plasma density through the use of a probing beam of fast atoms is convenient and useful even in the absence of information on the electron temperature.

In the analysis of the data one simplification is usually made: the formation of negative hydrogen ions is usually neglected. This procedure is also completely valid because of the small value of the corresponding cross sections for particle energies of the order of 10^4 eV.

We now consider several actual applications of plasma probing by atomic beams.

1. A beam of neutral hydrogen atoms is directed perpendic-

ularly to the lines of force of a magnetic field that contain the plasma. Each loss of an electron by a fast atom leads to the loss of the particle from the beam because of the deflection of the protons in the strong magnetic field. The change in the density of particles in the beam is described by the equation

$$\frac{dN_0}{dl} = -N_0 n_0 \sigma_{01} - N_0 n_i \sigma_{10}, \tag{8.5}$$

whence

$$N_0 = N_0(0)\, e^{-(n_0 \sigma_{01} + n_i \sigma_{10})}. \tag{8.6}$$

The density of neutral atoms in the high-temperature plasma is generally smaller than the density of ions. Assuming also that $\sigma_{10} \gg \sigma_{01}$, we can omit the first term in the exponent in Eq. (8.6). As a result we obtain a unique relation between the attenuation of the beam of neutral atoms and the product nl. A curve showing the attenuation of a beam as a function of nl is shown in Fig. 8.1. The absence of any data on the relative density of n_0 and n_i complicates the interpretation of the measured results considerably. In this case we only have a unique relation for the quantity $(n_i \sigma_{10} + n_0 \sigma_{01})$ and additional information is required in order to determine n_i and n_0 individually.

2. We now consider the probing of a plasma by an atomic beam in the absence of a magnetic field. Similar conditions hold for the use of a probing beam directed along the lines of force of the magnetic field. These conditions obtain, for example, in the transmission of a beam of neutral atoms along the axis of a mirror machine.

The change in the charge state of the beam with $\mathbf{H} = 0$ is given by the expression

$$\frac{dN_0}{dl} = n_0 N_i \sigma_{10} - n_0 N_0 \sigma_{01} - n_i N_0 \sigma_{10}. \tag{8.7}$$

Here

$$N_i + N_0 = N(0).$$

The density of neutral atoms in a beam which originally consists only of neutrals, and which moves through a distance l in a plasma characterized by the parameters n_0 and n_i, is

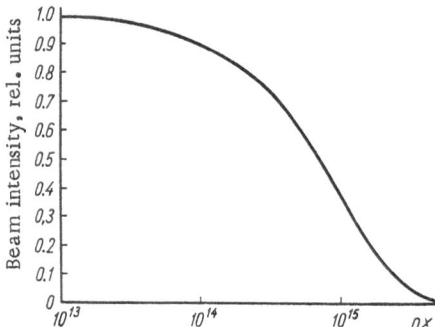

Fig. 8.1. Curve showing the attenuation of a
beam of neutral atoms of hydrogen in a fully
ionized plasma when the probing beam is directed perpendicularly to the lines of force of
the magnetic field.

$$N_0 = N_0\,(0) \left[\frac{n_0\,\sigma_{10}}{n_0\,\sigma_{10} + n_0\,\sigma_{01} + n_i\,\sigma_{10}} + \frac{n_0\,\sigma_{01} + n_i\,\sigma_{10}}{n_0\,\sigma_{10} + n_0\,\sigma_{01} + n_i\,\sigma_{10}}\, e^{-\,(n_0\,\sigma_{10} + n_0\,\sigma_{01} + n_i\,\sigma_{10})\,l} \right].$$

$$(8.8)$$

If the original beam consists only of protons, the expression
for the concentration of protons in the beam after a distance l in
the plasma is

$$N_i = N_i\,(0) \left[\frac{n_0\,\sigma_{01} + n_i\,\sigma_{10}}{n_0\,\sigma_{10} + n_0\,\sigma_{01} + n_i\,\sigma_{10}} + \right.$$
$$\left. + \frac{n_0\,\sigma_{10}}{n_0\,\sigma_{10} + n_0\,\sigma_{01} + n_i\,\sigma_{10}}\, e^{-\,(n_0\sigma_{10} + n_0\sigma_{01} + n_i\,\sigma_{10})\,l} \right]. \qquad (8.9)$$

Both of these expressions are rather complicated, but can be
simplified considerably if the relative density of neutral atoms in
the plasma is small; in this way they become convenient for the
analysis of experimental results.

A great deal of interest also attaches to methods of probing
with an atomic beam in measurements of the distribution of a neutral gas that changes with time. This problem is especially important in plasma devices in which pulsed gas emission is used,
for example, in electrodynamic plasma injectors [14]. If a dose of
gas is introduced into a magnetic field the measurements can only
be carried out with a neutral beam; the results can be analyzed

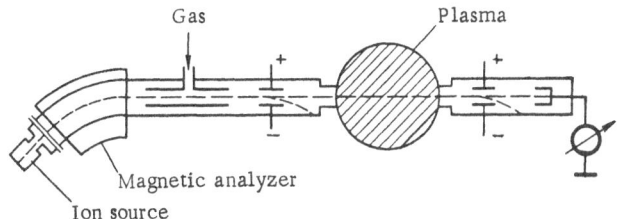

Fig. 8.2. Diagram showing probing of a plasma by a beam of
neutral particles.

using Eq. (8.6), taking $n_i = 0$. In the absence of a magnetic field
the probing can be carried out using a proton beam. The relation
between the gas density and the beam attenuation can be obtained
directly from Eq. (8.9) which, in the present case, assumes the
form

$$N_i = N_i(0) \frac{\sigma_{01}}{\sigma_{01} + \sigma_{10}} \left[1 + \frac{\sigma_{10}}{\sigma_{01}} e^{-n_0(\sigma_{10} + \sigma_{01}) l} \right]. \qquad (8.10)$$

In writing the formulas for the attenuation of an atomic beam
transmitted through a plasma we have assumed that the plasma
density is uniform and that the plasma has well-defined boundaries.
In general this assumption is not valid and the quantity nl must be
replaced by $\int n dl$. In writing the expression for nl without the integral we are to understand the product of some effective dimension multiplied by a mean plasma density. The method described
here is also used for measuring electron temperature; in this case
the plasma is probed by a beam of atoms with high cross sections
for ionization by electron impact (cf., for example, [6]).

In each individual case it is necessary to take account of the
charge states of the beam that are possible under the given conditions and to use the appropriate cross sections.

We shall now consider briefly a typical experimental arrangement used in the probing of a plasma by a beam of fast
neutral atoms (Fig. 8.2). A beam of hydrogen ions from a source
passes through an accelerating potential and is directed into a
magnetic analyzer. The analyzer is used to isolate the proton
part of the beam. The protons then pass through a so-called gas
target, which is a thin tube, in the center part of which there is an

aperture for the admission of gas. The amount of gas that is admitted and the length of the tube are chosen in such a way that the probability for charge exchange on the target is close to unity for the appropriate vacuum in the chamber. In passing through the target the protons capture electrons and are converted into neutral atoms. In order to eliminate protons which pass through the gas target without being converted into neutral atoms, there is an electric field which is produced by a plane condenser in the path of the beam; the direction of this field is perpendicular to the velocity of the particles in the beam. After passing through the plasma the beam is detected. The principle of operation of the detector is based on secondary emission of electrons from a metal surface bombarded by fast neutral atoms. This detector records charged particles as well as neutral particles; hence, in order to isolate the neutral component of the beam, directly in front of the detector there is a region of magnetic field or electric field in which the proton trajectory is manipulated so that protons are ejected from the beam. By measuring the intensity of the beam of neutral atoms in the presence of the plasma and without plasma it is possible to obtain the attenuation $N_0(l)/N_0(O)$ which is needed for determining $\int n dl$.

In analyzing this curve it should be kept in mind that the shape of the curve can be affected by the presence of long-lived metastable hydrogen atoms in the beam. The cross sections for the stripping of highly excited metastable atoms is very different from the corresponding cross section for hydrogen atoms in the ground state. The presence of metastable atoms in the beam is controlled by destroying the metastable state in an electric field of 30-40 kV/cm. It is also possible to destroy the metastable state by using the Lorentz electric field $\frac{1}{c}$ [v, H], which arises in the motion of the beam through the magnetic field.

A basic difficulty in the measurements lies in the elimination of the background of secondary emission due to the radiation from the plasma itself. The background can be reduced by good collimation of the beam and by removing the detector to large distances from the plasma. A significant increase in the signal-to-noise ratio can be obtained by modulating the density of the beam from the source and detecting the detector current in a tuned amplifier which is tuned to the modulation frequency.

The density of hydrogen atoms, neutral or ionized, can be determined with the required accuracy only when the loading of the hydrogen plasma by impurities due to other atoms or ions is small. If this is not the case it is necessary to take account of charge exchange of the beam on atoms of the impurities; the cross sections of these impurity atoms can be very different than for hydrogen.

Neutral particles with thermal energies can also be used for probing a plasma [15]. Because of the low velocity of these particles the basic mechanism for the change in the charge state of the beam is usually ionization by electron impact. This is particularly the case in probing by a beam of atoms which do not appear in the makeup of the plasma itself, in which case there is no resonance charge exchange.

A flux of particles with thermal velocities, or, as it is sometimes called, a molecular beam, is formed by evaporation in vacuum from the solid phase with subsequent effusion of the gas molecules through special channels. The distribution of particle velocities in a molecular beam is described by

$$I\left(v\right) = \frac{2I_0}{\sqrt{\dfrac{2kT}{M}}}\, v^3\, e^{-\frac{Mv^2}{2kT}}. \tag{8.11}$$

The detection and measurement of molecular beams is accomplished by means of detectors that operate on the basis of various physical principles. The most popular method of detection is based on the surface ionization of atoms which strike a tungsten plate heated to temperatures of 1300–1500°. This method of detection is widely used for detection of molecular beams of alkali metals, which are characterized by low ionization potentials. A beam of sodium or a beam of lithium can be detected on an oxidized tungsten surface. The sensitivity of the method of detection based on atomic ionization can be improved significantly by using a multiplier with an open cathode in conjunction with an acceleration space. Photoelectric detection essentially yields the possibility of detecting individual ionized atoms.

A method which is characterized by much lower sensitivity, but which is much simpler, is based on the measurement of the

electrical conductivity or the transparency of a condensed film.
Methods of this kind have been used in work with molecular beams
of the noble metals. The technique for producing the molecular
beams and for measuring them has been developed over many
years and the state of research in this field has been described in
detail in a number of reviews (cf., for example, [16]). Since molecular beams have not been widely used in plasma diagnostics it
would not be useful to discuss this method in detail at this point.
Measurements of the density of deuterium in the plasma have also
been made with a tritium beam at energies of 150-200 keV [17].
The application of this method is based on the assumption that the
plasma in the device is formed by the ionization of the heavy isotope deuterium, rather than ordinary hydrogen. The maximum
cross section for the D(T, n) He^4 reaction is $5 \cdot 10^{-24}$ cm^{-2} and is
achieved for a tritium energy of 160 keV. The neutron yield (or
the yield of α particles with energies 3.5 MeV) is proportional
to $\int nd\mathit{l}$.

Depending on the requirements imposed on a particular experiment it is possible to measure the deuterium concentration by
recording the α particles or the neutrons. By using a collimator
to isolate the α particles formed in a given region occupied by the
plasma it is possible to obtain the density distribution function
in the path of the tritium beam. An experiment of this kind is
more exact but it is somewhat complicated by the emission of α
particles from the discharge chamber. It is much simpler to detect the output of neutrons with a neutron counter located outside
the discharge chamber. The spatial distribution of deuterium
density in this experimental setup can be determined by moving
the tritium beam over the cross section of the plasma. For a
beam current of 1 mA the lower limit in the possible measure-
ment of density by a tritium beam is approximately $\int nd\mathit{l} = 10^{15}$ cm^{-2}.

The primary error in the determination of plasma density
by this method comes from the background of neutron radiation
that is produced when the beam strikes the surface of the chamber
in which there is absorbed deuterium. A particularly large contribution to the background comes from neutron irradiation from
the plate on which the tritium beam is brought to rest after passing through the plasma. This background can be reduced by heating the detection plate, which is made of stainless steel, to temperatures of 1000°C.

An important feature of operation with a tritium beam is the maximum possible limitation of the tritium yield in the ion source. This can be achieved most easily through the use of a pulsed ion source with tritium-saturated buttons [18, 19]. Unfortunately, the method based on the tritium beam is not always useful since it requires rather complicated apparatus and requires the observation of special safety conditions for operation with tritium.

§ 8.3. Mass Spectrometer Method

for Studying Fast Particles in a Plasma

In 1952 hard radiation was detected in a powerful pulsed discharge [20, 21], this radiation being caused by nonequilibrium groups of fast particles in the plasma; shortly afterward particles with high energies were found in other devices intended for obtaining high-temperature plasmas. In spite of the rather large amount of published work devoted to the investigation of this effect, the origin of the nonequilibrium group of fast particles remains unexplained. Usually these particles are attributed to the presence of a plasma instability. If this assertion is correct then the superthermal particles that arise in a plasma may evidently be taken as an indication of an instability of a plasma in a magnetic field in a given configuration. It will be evident that the investigation of fast particles that arise in a plasma is still of interest to physicists who are concerned with controlled thermonuclear research.

Fast particles in a plasma can usually be observed easily by virtue of secondary effects that arise, leading to the production of hard radiation which is capable of penetrating the walls of the discharge chamber. For example, x-ray radiation is produced in the retardation of fast electrons in the plasma or at the walls of the discharge chamber. The x-ray photons formed by the retarded electrons, with energies reaching tens of kiloelectron volts, easily penetrate the walls of the discharge chamber. In this case, conventional methods used in radiation physics can be used to determine the energy and intensity of the x-ray beam. In choosing a method of measurement it is necessary to avoid the use of instrumentation that is sensitive to the electromagnetic fields that are frequently found around operating thermonuclear devices. In the investigation of x-ray radiation at energies below several kiloelectron volts additional difficulties arise that are associated with the transmission of the radiation through the walls of the chamber. Soft x-ray radiation can be transmitted through a window fabricated from aluminum or beryllium. In order to obtain transmission of photons with energy of approximately 1 keV it is necessary to use beryllium foils with thickness as small as 1 μ.

The presence of fast deuterons in a plasma can be observed from the neutron radiation that is produced as a result of D—D reactions. A detailed review of methods that are used for the detection of neutrons and x-ray radiation from a gas discharge is contained in the following section.

Information on fast particles in a plasma, which is obtained by investigating the hard radiation, frequently does not yield the required accuracy. Thus, it is difficult to obtain definite results from such measurements without some information as to the nature of the spectrum of the fast particles or without some knowledge concerning the region in which the acceleration occurs. In many cases there is no possibility of detecting fast particles through secondary processes. Hence, it becomes necessary to use methods that detect these fast particles by determining the specific charge Ze/M and the energy. The mass spectrometer is a suitable instrument for this purpose.

At the present time, in nuclear physics wide use is made of complicated mass-spectrometer devices which have high resolving power and which can be used to measure the mass of particles with tremendous accuracy. Work in plasma physics does not require this high accuracy and hence there is no need to use complicated spectral devices. As a rule, in plasma work it is sufficient to make use of rather elementary mass spectrometers which do not differ very much from those that were used in atomic physics some ten years ago. However, in the mass-spectrometer analysis of plasma there are certain specific problems that arise and questions related to this topic are discussed in the present section.

As a rule, mass-spectrometer methods used for the detection of fast electrons are not found suitable for the investigation of fast ions. The point here is that in practice, even in plasmas formed by the ionization of a spectrally pure gas, there are always ions with different charge and, frequently, different mass. The energies of fast electrons can be measured in a simpler way since the problem is not complicated by an unknown specific charge.

The extraction of a beam of fast electrons from a discharge chamber is usually accomplished by means of small windows made from aluminum foil or beryllium foil. In the investigation of electrons with energies greater than twenty kiloelectron volts the extraction of the beam can be realized through a thin slit iris. In this case, the capacity of the vacuum pumps that maintain the vacuums in the chamber of the mass spectrometer must provide the required pressure differential. The pressure of the residual gas in the chamber of the mass spectrometer must be small in order to eliminate scattering of electrons by molecules of the residual gas. In a vacuum of about 10^{-5} mm Hg the scattering of electrons has essentially no effect on the results of mass spectrometer investigations of fast electrons from a plasma.

The vacuum chamber of the mass spectrometer is located between the poles of an electromagnet which is supplied by a stabilized dc source. In passing through the collimator in the region of the magnetic field the electrons experience a Lorentz force $\frac{e}{c}$ [v, H]; as a result the beam is formed into a fan-like structure determined by the initial distribution of particle velocities. In order to increase the "optical power" of the device the slit in the collimator must be located as closely as possible to the region of magnetic field; on the other hand, care must be taken to see that the fringing magnetic field does not displace the beam in the collimator itself since this effect can lead to distortions in the measured results.

When the intensity of the beam is inadequate to expose a photographic film in a device with high resolution and it is not necessary to make a detailed investigation of the electron spectrum, use can be made of a simple spectrometer that employs direct deflection [22]. In this device the high-energy electrons pass through the magnetic field and are deflected through small angles; the dispersion of the device is small for high energies while the transmission is large. In spite of the extreme simplicity of this mass spectrometer the results that are obtained are frequently very useful, especially when these measurements are used in conjunction with other methods.

In determining the electron energy from a spectrogram taken on such a spectrometer it is recommended that the electron trajectories be plotted and that the device be calibrated by means of a beam of electrons of known energy. In plotting the electron trajectories important errors can arise from effects due to the scattering field if one considers large distances from the magnet. It will be evident that direct calibration of the device yields the most reliable and exact data. If a discharge contains a flux of electrons of known energy in addition to the electrons being investigated as is the case, for example, in the striations of a low-pressure pulsed discharge, these electrons can be used as a reference. This same kind of electron flux can be used to monitor the position of the film in the device by exposing it in the absence of the magnetic deflection field.

More detailed information on the electron energy spectrum in a plasma can be obtained with a so-called 180° spectrum analyzer [23]. This spectrum analyzer is convenient in that it can be used for rough determinations of the electron energy by measuring the magnetic field in the chamber, making use of the familiar relation

$$\rho = \frac{mvc}{eH}, \quad \text{or} \quad \rho \text{ (cm)} = 3.3 \sqrt{W} \text{ eV}, \tag{8.12}$$

and computing the position of the image of the slit corresponding to a given electron velocity. In the solution of certain problems it is also possible to use devices that are based on the deflection of charged particles in an electrostatic field.

The complicated nature of the spectral sensitivity of photographic films makes it difficult to determine the electron spectrum on the basis of the distribution of blackening density on a film. Hence, when it is necessary a given film can be calibrated by means of an electron beam of known intensity. In the investigation of electron energies of several tens of kilovolts it is found convenient to use photographic films that are sensitive to x-ray radiation, especially those used for the detection of soft x-ray radiation. Other films, in particular sensitive movie film, are also used.

In the investigation of ions which are emitted from a thermonuclear device the primary problem, in contrast with that which arises in electron work, is the determination of mass. The importance of this information is due to the fact that the presence of heavy ions leads to a rapid cooling of the plasma as a result of radiation in the ultraviolet region and the soft x-ray region. If the subject of the investigation is a relatively cold plasma, whose ion energies do not exceed hundreds of electron volts, the mass composition of the ions can be analyzed by any mass spectrometer in which the ions pass through an accelerating potential. The width of the lines on the film will be determined by the energy distribution of these ions in the plasma. A mass analysis is difficult if the ion energy extends to tens and hundreds of electron volts, as is the case in work on nonequilibrium particle groups in powerful pulsed discharges. At these energies successful use can be made of the parabola method first introduced by Thomson; this method is now being used more and more in work on controlled thermonuclear reactions (cf., for example, [24, 25]).

In the parabolic method for the analysis of ions by charge, mass, and energy, a beam of fast charged particles (isolated by an aperture in the wall of the discharge chamber) enters a space in which there are uniform electric and magnetic fields whose lines of force are parallel to each other. Along the path of the beam there is a collimator which is designed to keep the scattered light from the device from reaching the photographic film. When the ions enter the region of electric and magnetic fields the ion trajectories are curved. The magnetic field deflects the charged particles in a direction perpendicular to the magnetic lines of force. The magnetic displacement of the beam, as observed on the photographic film under the assumption that small angle deflection obtains, is given by

$$x = \frac{ZeHl^* h}{Mv} , \tag{8.13}$$

where l^* is the distance between the magnet and the film and h is the dimension of the region occupied by the magnetic field.

In the electric field the charged particles move along the parabola which lies in the same plane as the trajectory of the particle in the absence of an electric field. This plane is parallel to the lines of force of the electric field. Under the effect of the electric field the beam will be displaced at right angles to the direction of displacement due to the magnetic field. The magnitude of the beam displacement due to the electric field is

$$y = \frac{ZeEl^* h}{Mv^2}. \tag{8.14}$$

By eliminating the velocity from (8.13) and (8.14) we obtain the equation of a parabola whose shape is a function of the ratio Ze/M for given electric and magnetic fields. Thus, the photographic film exposed to an ion beam that has passed through a region of crossed fields will show a family of parabolas, each of which corresponds to an ion having a given ratio Ze/M; the different points on the parabola then correspond to ions of different energy. The analysis of ions in crossed fields is especially useful for work with plasmas in which there are impurities of heavy atoms that appear at various stages of ionization. The parabola method makes it possible to obtain an energy distribution curve for each ion species. The validity of the identification of any given parabola can easily be verified by exposing the film'to ions of known mass. These supplementary measurements can be carried out by filling the discharge chamber with spectrally pure gases.

By measuring the blackening of the film along each of the parabolas and using the spectral sensitivity curve for the photographic film that is being used it is possible, in principle, to obtain detailed information on the energy distribution of each ion species. The region of energy that is of interest for thermonuclear research is in the tens of kilovolts and in some cases hundreds of kilovolts. Unfortunately, this is precisely the range in which there are almost no experimental data on the sensitivity of photographic films. The only work in which the spectral sensitivity of photographic films has been investigated systematically for various ions does not allow a high accuracy extrapolation to other photographic materials [26].

The photographic materials that are available at the present time have low sensitivity to low-energy particles. Thus, the detection of ions with energies in the hundreds of electron volts involves a number of difficult problems. In making use of the parabola method in such cases it is usually found useful to use an accelerating gap in which a known potential difference is applied. In experiments with plasma injectors carried out at the Khar'kov Physicotechnical Institute [22] the sensitivity of the parabola method has been increased by using an accelerating gap. Before entering the mass spectrometer the ions are accelerated by the same potential

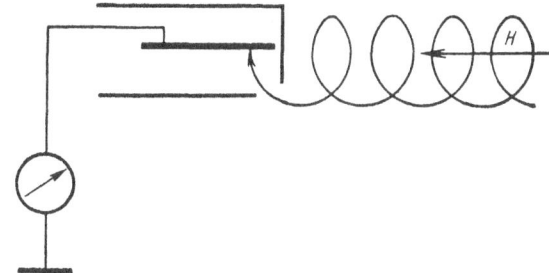

Fig. 8.3. Diagram showing an instrument for measuring
particle energy in a trap.

difference. As a result, all of the parabolas on the photographic
film exhibit greater sensitivity but do not start from zero; rather
they start from a point displaced from zero by an amount corresponding to the accelerating voltage.

In using the mass-spectrometer method one should be aware
that the results obtained by this method can sometimes be distorted
by electric fields due to the walls; this effect must be taken into
account in the analysis of measurements.

In estimating the energies of ions in a plasma we are not necessarily confined to an external magnetic field. It is also possible to
use the magnetic field within the device itself (cf., for example, [27]).
As an example we consider measurements in the Orekh system,
which has a minimum-B field. Ion analysis is carried out through
the use of the region of uniform magnetic field using magnetic slits.
The ion collector is a thin metal plate located in a plane parallel
to the lines of force of the magnetic field. The collector is located
within a metal box. The side of the box turned toward the plasma
does not admit particles that move along the lines of force into
the collector. However, particles moving along a helical trajectory can enter the box through a slit cut into one of the side walls.
The arrangement of the collector and slits and the particle trajectories are shown schematically in Fig. 8.3. If the collector is far
from the side wall containing the slit, at a distance equal to 2ρ,
then no particle with a trajectory characterized by a radius of
curvature smaller than ρ can enter.

By locating the collector at different distances from the slit
and measuring the current strength at the collector as a function

of distance it is possible to estimate the ion energy and to determine the change in this energy as a function of time during the
lifetime of the plasma in the trap. At the present time an analysis
of this method is not yet available and hence the value of the energy
of the ions which is obtained in this way is to be regarded as approximate. If the plasma density is reasonably large it will be
found that the measured results are affected by the space charge
field due to the plasma electrons. This field inhibits the loss of
ions from the plasma across the magnetic field.

Investigation of the energy distribution of particles that leave
the discharge is of considerable interest only when these particles
are representivative of the plasma being investigated or when they
arise as a consequence of processes that characterize the behavior of the plasma under actual conditions. Charged particles
that escape from a system can have an energy spectrum that
differs considerably from that of the particles in the plasma.
Thus, measurements of the energy distribution of such particles
does not always give useful information concerning the parameters
of the plasma contained in the trap. More representative, in this
respect, are the neutral atoms formed by charge exchange of
fast plasma ions, primarily on the residual gas. The fast neutral
particles formed by charge exchange escape from the magnetic
field region at any angle with respect to the lines of force. The
energy distribution of these particles has been investigated in
greater detail on the Alpha device through the use of a mass spectrometer [28-30] and on the PR-5 device through the use of thin
transmission films [31, 32].

A diagram of the apparatus used for investigating charge-
exchange particles in the Alpha device is shown in Fig. 8.4.
The neutral particles are detected by the most popular method,
namely, emission of secondary electrons from a metal surface.

The beam of neutral particles and charged particles is ex-
tracted through an aperture in the wall of the discharge chamber
and pass through the electric field produced by a condenser K.
After the charged component is removed from the beam the neutral particles enter a gas-filled chamber CE (charge exchange
chamber) where, as a consequence of stripping, fast neutral atoms
loose electrons and are converted into singly charged ions with
essentially no loss of kinetic energy. The ions formed by charge

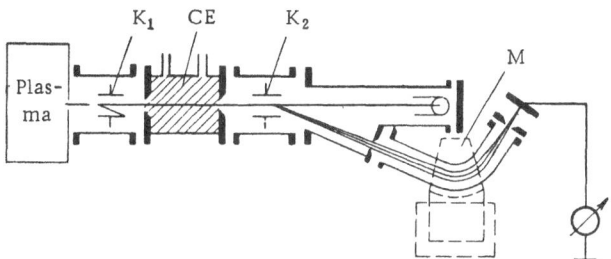

Fig. 8.4. Device for studying fast neutral particles emitted from a discharge: K_1 and K_2 are plates used to produce the magnetic field; CE is the gas-filled charge-exchange chamber; M is the magnetic analyzer.

exchange enter the electrostatic analyzer, which isolates a narrow energy range. The ions pass through the electrostatic analyzer and enter the magnetic field, in which a mass analysis of the beam is carried out. This analysis is possible because of the low probability of loss of two electrons from an atom in the charge-exchange chamber. The ion beam from the charge exchanger consists primarily of singly charged ions.

The efficiency of detection for the neutral atoms is determined by a calibrated beam with known parameters. For this purpose a special instrument has been developed; a description of this device is given in [33].

These measurements, as expected, have shown that the spectrum of neutral particles consists not only of hydrogen atoms, but also of a large number of impurity atoms, in particular, carbon and nitrogen. The energy distribution curves fall off rapidly at high energies. When the energy is increased from 1000 to 4000 eV, the number of particles per energy interval is reduced by approximately four orders of magnitude.

§ 8.4.   Calorimetric Methods
of Plasma Diagnostics

Almost all the investigations of plasma behavior in various devices are intended for the investigation of high-temperature plasma. Special interest attaches to information concerning the magnitude and distribution of the heat flux to the walls of the vacuum chamber. Depending on the nature of the experiment, usually one encounters at least one of the three following problems:

1. The determination of the distribution curve for the heat flux to the walls in relative units.

2. The determination of the energy evolved at a given region of the surface (absolute measurements).

3. The determination of the time behavior of the heat flux.

The first problem is the simplest. To solve this problem, generally thermal detection elements are located in the walls of the chamber, these elements being made of metal sheets. To each of these thermal elements is attached a separate thermocouple. The choice of the thickness of the thermal detection element depends on the sensitivity of the measurement instrument and on the amount of energy being absorbed by the thermal detection device. The thermal detection devices are usually made of stainless steel. If, per unit surface of the thermal detection element, there is evolved an amount of heat given by Q calories, the increase in temperature is given by

$$\Delta \theta = \frac{Q}{dch} (1 - \gamma). \tag{8.15}$$

Here d is the density, c is the heat capacity, and h is the thickness of the thermal detection device, while γ is the coefficient of reflection of the plasma from the surface of the thermal detection device. Usually the coefficient γ is of the order of ten per cent. Assuming that the fraction of the energy intercepted by the detectors is independent of Q (this assumption holds over a wide range of values of the heat flux) and assuming that the thermal detection devices that record the temperature are located at various points, it is possible to determine the heat distribution at the wall in relative units.

It will be evident that heating of the thermal detection device is due not only to the plasma that strikes its surface, but also to radiation and fast particles, the fast particles arising from charge exchange. The possibility of determining the contribution of each of these processes without a detailed analysis depends on the nature of the objects being investigated. Thus, the existence of an anisotropy in the heat flux from the plasma almost always indicates the transfer of energy to the walls by charged particles. In particular, the presence of a clearly defined peak in the distribution at a point at which the lines of force intercept the walls cannot be a consequence of radiation or a flux of neutral particles, since both of these effects would be independent of the field near the thermal detection device. This is the situation in measurements of the heat flux to a wall of a chamber used in a cusp device [34]. Here the clearly defined peak observed in the region of the edges of the magnetic slits indicates the predominant loss of plasma along the lines of force. Detection of fast neutral particles and radiation from a plasma can sometimes be carried out by means of various filters. An interesting possibility that can be used to determine the contribution of neutral particles separately from the ultraviolet radiation has been considered in [35, 36], in which gas filters were used.

In order to determine the absolute value of the energy transported from the plasma to a given section of the wall surface of the vacuum chamber it is necessary to know either the entire incident energy or the reflection coefficient γ. In the first case the problem is solved most simply by means of a calorimeter with a processed surface. This calorimeter is fabricated in the form of a deep cylinder furnished with partitions that are perpendicular to the bottom. Multiple reflection from the partitions leads to complete absorption of the energy by the calorimeter. In certain cases the reflection coefficient of the plasma does not go appreciably beyond 50% and in order to determine the energy transported by the plasma with an accuracy of several per cent it is sufficient to "develop" the surface of the thermal detection element by a factor less than ten. It should be kept in mind that in the presence of a magnetic field the reflection occurs primarily in the direction of the lines of force. Consequently, when the axis of the calorimeter is oriented along the magnetic field the distance between the partitions must be as small as possible.

The time variation of the energy transported by the plasma is usually investigated by means of special bolometers. High sensitivity and good resolution time are exhibited by a three-layer bolometer which has been developed by Gorelik [37–39]. On the electrolytically oxidized side of an aluminum foil is vacuum

deposited a thermal resistance alloy of bismuth and lead. The
thermal resistance, fabricated from 0.6% lead and 99.4% bismuth,
has a negative temperature coefficient of 0.3% per degree. Applying a voltage of 10 V to the thermal resistance it is found that
the resistance is approximately 5 kΩ and it is possible to obtain
a sensitivity of approximately 10^{-4} J/cm^2 with a resolving time of
1 μsec. These characteristics are exhibited by a bolometer fabricated of aluminum foil 6 μ in thickness with an active layer
that is 1-2 μ in thickness.

The resolution time of the bolometer is determined by the
rate of equilibration of the temperature in the direction perpendicular to the surface of the foil. An important characteristic
of the device is the time of equilibration of the temperature along
the surface of the foil. In order to minimize the distortions of
the signal being investigated this time must be significantly greater
than the characteristic time for the process being investigated. In
the bolometer being considered here the temperature equilibration time along the foil is greater than 10^{-3} sec. If the length of
the heat pulse is greater than the resolution time of the bolometer,
but smaller than the equilibration time along the foil, the change
in the temperature of the thermal resistance at time t is given by
the expression

$$\Delta\theta\,(t) = \int_0^t \frac{1}{dch} \cdot \frac{dQ}{dt}\,. \tag{8.16}$$

A curve showing the dependence of the energy carried by the
plasma flux as a function of time is obtained by graphical (or electronic) differentiation of the signal from the bolometer, which is
recorded by means of an oscilloscope.

The greatest sensitivity in devices of this kind is obtained in
a three-layer bolometer, which is fabricated from a colloidal
film. On the film side that faces the plasma there is deposited
a layer of silver approximately 1 μ thick. On the opposite side is
deposited a thermal resistance alloy of bismuth and lead. This
construction exhibits high sensitivity; the layer of silver plays the
role of an electrostatic shield which attenuates the background
noise to a great extent. The bismuth−lead bolometer can work in
the presence of either a fixed or quasistatic magnetic field. The

use of a colloidal base bolometer for calorimeter measurements on the magnetic device Orekh made it possible to measure heat fluxes as low as 10^{-5} J/cm^2 [39].

We wish to consider briefly at this point an original method for the detection of short heat pulses which has not received wide application in plasma physics. This method is based on the detection of radiation from a thin tungsten foil which is heated by a plasmoid [40]. When it is near room temperature the thermal radiation of the foil lies in the far infrared, which is not suitable for detection by a photomultiplier; in this case the device has a high threshold. In order to increase the sensitivity the foil is heated by an electric current to a temperature of several hundreds of degrees Kelvin, in which case a small change in temperature due to the heating by the plasma is easily recorded by the photomultiplier.

The radiation of the foil is measured by a photomultiplier which is located behind the foil in such a way that the plasma radiation cannot reach the photocathode directly. A basic advantage of the method lies in the absence of electrical contact between the foil, which is in contact with the plasma, and the photomultiplier; this feature makes it possible to eliminate leads to a measurement device.

§ 8.5. Detection of Hard Plasma Radiation

The measurement of the energy and intensity of hard radiation from a plasma becomes more and more important from year to year. At the beginning of thermonuclear research, interest in the study of hard radiation in a plasma was stimulated only by the necessity of obtaining information on the development of instabilities, which led to the appearance of a nonequilibrium groups of fast particles [21, 22]. In recent years devices have been built in which plasma has been contained at electron temperatures of 10^4-10^5 eV; consequently, the investigation of the shape of the spectrum and the intensity of the x-ray radiation is important for the determination of the electron temperature and density. Significant interest also attaches to neutron measurements, since the parameters that can be achieved with present-day plasmas are such that the measurement of neutron fluxes in a deuterium plasma will evidently provide an independent determination of the ion temperature in the near future.

Methods of investigating the hard radiation have been successfully developed over the years in nuclear physics and an ex-
haustive description of these methods can be found in many books
on the subject. Nevertheless, in the interest of obtaining a complete presentation, it will be useful here to dwell briefly on those
methods which are most appropriate in plasma physics. In contrast with contemporary nuclear physics, in plasma physics the
requirements on the accuracy of measurements are not very high;
hence the simplest and easiest methods have received wide acceptance. For this reason in the present section, as in our analysis of mass-spectrometer methods, we shall consider only the
simplest methods and those which have received widest use in experimental work in plasma physics.

The Induced-Radioactivity Method. This procedure is used in the observation and measurement of neutron
pulses of 10^5-10^6 neutrons and higher, under conditions in which the
detection of neutrons by other methods would be difficult because
of the background of electromagnetic noise and when it is not
necessary to investigate the shape of the neutron pulse. The single
requirement that pertains to the induced-activity method is that
the length of the neutron pulse be short compared with the half-
life of the active isotope. This feature makes it possible to use
apparatus which records the activity of a target after the working
cycle of the apparatus is completed and electromagnetic noise is
no longer a problem. The method of induced activity was used
in the discovery of neutron radiation from a pulsed discharge in
deuterium at high current [20]. A schematic diagram of the detector which is based on the detection of β radiation that arises
in the capture of slow neutrons by silver isotopes is shown in
Fig. 8.5. The neutrons are slowed down by multiple scattering
in a paraffin block and are captured by silver nuclei in the target,
which is surrounded by a β counter. Silver has two isotopes, Ag^{107}
and Ag^{109}, in approximately equal amounts. The cross section for
capture of thermal neutrons in these isotopes is 44 and 110 b,
respectively. Capture results in the formation of the β radia-
tors Ag^{108} and Ag^{110} with half-lives of 2.4 min and 24.5 sec, respectively. Thus, the silver target makes it possible to record pulses
with lengths up to several seconds. If the neutron pulse has a
short lifetime and if the duty cycle of the device does not exceed
several minutes, detection of very small neutron bursts can be

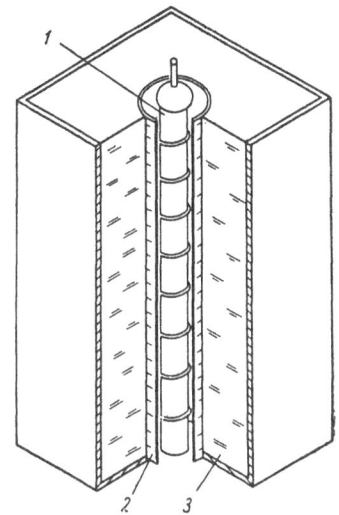

Fig. 8.5. Neutron detector that makes use of induced activity of a silver foil. 1) β counter; 2) silver foil; 3) paraffin block.

accomplished through the activity that arises in the capture of a neutron by the manganese isotope Mn^{55} [41]. The half-life for the β-active manganese is 2.59 h. The activization of the target requires several pulses. This method is known as the induced manganese method in nuclear physics.

Since the efficiency for the detection of neutrons by the induced-activity method is rather difficult to compute, in carrying out absolute measurements of the neutron yield the apparatus is usually calibrated by a known neutron source.

Scintillation Detector. One of the most popular methods for the investigation of hard radiation is based on the use of a scintillation detector. This is used both for the investigation of the time characteristics of the radiation as well as for the determination of the energy spectrum. Bursts of visible light that arise in the scintillating crystal under the effect of neutrons or x-ray photons are recorded by a photomultiplier. Pulses from the output of the photomultiplier are observed on the screen of an oscilloscope. The neutrons are detected by means of organic crystals which have the further advantage of good resolution time. For example, the emission time for anthracene is $3 \cdot 10^{-8}$ sec; for stilbene the emission time is $6 \cdot 10^{-9}$ sec. The luminescence in these crystals is caused by recoil protons. In experiments that require measurements of the hard x-ray radiation use is also made of heavy crystals (sodium iodide, cesium chloride). In these crystals the luminescence is produced by Compton electrons and photoelectrons. In heavy crystals the smallest emission time ($\sim 2 \cdot 10^{-7}$ sec) is found in sodium iodide activated with thalium. The detection of soft x-ray radiation which passes through the discharge chamber through a window made from vacuum-deposited aluminum or beryllium foils is best realized by means of thin scintillation plates or films [42]. If various processes simultaneously give rise to soft and hard x-ray radiation, the use of this scintillator reduces the efficiency of detection sharply for hard photons and, in this way, reduces the hard background.

The requirements imposed on a detector designed for experiments on the determination of the electron temperature by analysis of the shape of the x-ray spectrum are much more severe than those that arise in the determination of the radiation pulse alone. The resolution time of the apparatus must allow the resolution of pulses due to individual photons. For these measurements use is made of good spectroscopic photomultipliers and crystals such that the apparatus width is much smaller than the spectral range being investigated. A Cs^{137} source is frequently used as a standard for calibration. A pulsed oscilloscope is used most frequently for measurements of the amplitudes of pulses at the output of the photomultiplier [43]. It will be obvious that the oscilloscope must also have adequate resolution time in order to resolve pulses due to individual photons. The radiation spectrum obtained in this way can then be used to determine electron temperature from the behavior of the short-wave region (cf. §5.1).

The measurements are complicated still further if it is necessary to determine the yield of the x-ray radiation of the plasma in absolute units, for example, when the density of a plasma is determined from the bremsstrahlung intensity. In carrying out these measurements it is necessary to use a spectroscopic crystal with known efficiency. The basic difficulty in measuring the intensity of x-ray radiation from a plasma lies in the elimination of the effect of the radiation background produced when fast electrons strike the wall of the chamber. The background can be reduced substantially through the use of collimators oriented in such a way that the field of view is sensitive only to a portion of the wall which is far from the plasma; this is done by means of a special cavity attached to the discharge chamber. The contribution of the radiation that arises from the wall can be estimated easily by locating at a section of the wall in the field of view of the detector a metal plate made from a material with a high atomic number (tantalum, molybdenum) and comparing the intensity of the recorded radiation in the presence of this target and without it [42]. If the basic contribution comes from wall radiation the x-ray photon yield will be changed in proportion to the atomic number of the target struck by the electrons. The fraction of the energy of electrons radiated in the form of a continuous spectrum in slowing down in a solid target is estimated from the formula

$$\eta \approx 1.2 \cdot 10^{-9} ZW, \qquad (8.17)$$

where Z is the atomic number of the target and W is the electron energy in electron volts.

It would appear that one of the most difficult problems that can be solved by means of a scintillation device is the detection of a weak x-ray radiation emitted by a gas discharge after an intense short burst of hard radiation [44]. Under these conditions, directly after the x-ray flash the output of the photomultiplier shows pulses corresponding to photon energies of about 10 keV; these photons are not related to processes in the plasma. These photons arise because of the emission delay of the scintillator. The long delayed emission is observed in CsI and NaI crystals as well as scintillators fabricated from stilbene [45, 46]. In stilbene a component with a long delayed emission of approximately 10^{-3} sec can have an integrated intensity of 0.1-1% of the intensity of the x-ray burst, causing an afterglow.

In order to eliminate the effect of the intense x-ray burst one can make use of a mechanical shutter which cuts off the radiation flux at the time of the burst. The resolution time of the shifter in a rotating chopper must, as a rule, be several tens of microseconds.

Pinhole Camera. The principle of the pinhole camera in the x-ray region of the spectrum is essentially the same as that of a standard pinhole camera such as is described in any textbook on optics. The sole difference lies in the fact that the walls are fabricated from lead and are opaque to x-ray radiation. In spite of its great simplicity, this device has made it possible to obtain a great deal of important information on the spatial distribution of the emission centers for x-ray photons. The use of a lead pinhole camera in experiments with powerful pulsed discharges made it possible to obtain the first data on the direction of acceleration of electrons produced in the development of instabilities [21]. In [47], in an investigation of a high-temperature plasma in a magnetic trap by means of a pinhole camera it was possible to determine the region occupied by the plasma containing the hot electrons. The photographs made in this work showed the absence of sources of x-ray radiation at the cylindrical chamber walls and this result appeared as the basic proof of the low loss of plasma across the magnetic field under the particular experimental conditions. The direct demonstration of the volume nature of the x-ray radiation also made it possible to determine the plasma density by using the expression for the intensity of the bremsstrahlung radiation.

Very frequently the intensity of the x-ray radiation is insufficient for the exposure of the film during the single discharge. It is possible to increase the sensitivity for photographic detection by using amplifying shields, for example, made from calcium tungstate. By placing the film between two amplifying shields it is possible to increase the sensitivity several times owing to the effect of fluorescence.

In the investigation of effects that accompany bursts of high-intensity x-ray radiation it is possible to realize a sweep of the image in time. Inasmuch as it is not possible, in the x-ray region, to apply the rotating mirror method, which is widely used in the optical region, in obtaining the time sweep it is more convenient to use a rotating drum or a disk with a film.

Analysis of the Tracks of Charged Particles in Nuclear Emulsions and in the Wilson Cloud Chamber [21, 48, 49]. In spite of a number of their shortcomings, nuclear emulsions have one important advantage as compared with the methods of studying hard radiation listed above. Along with the determination of the energy of the recoil protons, they provide the possibility of finding the direction of the deuterons produced as a result of D−D reactions if the reaction is due to the interaction of a group of accelerated deuterons with deuterons in the target or a low-temperature plasma. Even at low values of the deuteron energy, as compared with the output of the reaction, the difference in energy of neutrons emitted at angles of 0 and 180° with respect to the deuteron beam is found to be appreciable. The energy of neutrons emitted at these angles is determined in the following way:

$$W_n = \frac{M_n}{2}\left(\frac{W_D}{2M_D} + \frac{3}{2}\cdot\frac{Q}{M_n} \pm \sqrt{\frac{3W_D Q}{M_D M_n}}\right), \tag{8.18}$$

where Q is the reaction yield (Q = 3.25 MeV); W_D is the energy of the incoming deuteron; M_n and M_D are the masses of the neutron and deuteron. The plus sign corresponds to the angle 0° and the minus sign to the angle 180°.

At low deuteron energies the dependence is found to be rather sharp. Thus, for example, for energies of incoming deuterons up to 50 keV the energy of the neutrons emitted at an angle of 0° is 2.7 MeV and at 180° it is 2.2 MeV. The neutron energy in the system fixed in the center of mass is 2.45 MeV.

Physicists at the University of California, using nuclear emulsions, have obtained somewhat later, but independently of work carried out at the Institute for Atomic Energy, Academy of Sciences of the USSR, proof that the neutron radiation arising in the passage of intense current pulses through deuterium in straight discharge chambers is primarily not of thermonuclear origin. The emulsions were located along the axis of the discharge chamber close to the two electrodes. The developed emulsions revealed tracks of recoil protons whose directions formed small angles with the axis. In [48] the limiting angle is 10°. Measuring the length of these tracks, and in this way determining the energy of neutrons emitted along the axis in both directions, it was possible to show that the neutrons were produced by a beam of deuterons accelerated along the axis with characteristic energies of the order of 100 keV. The validity of the results was checked in a special experiment in which the emulsion was irradiated by neutrons from a deuterium target which was bombarded by a beam of deuterons of known energy. A Wilson cloud chamber can be used to measure the limiting energy of hard radiation. The use of a Wilson chamber is indicated in those cases in which the length of the radiation pulse is small and individual bursts cannot be resolved easily in the scintillation contour. Under these conditions the Wilson cloud chamber has an advantage as compared with electron-sensitive nuclear emulsions. The small sensitivity time for a chamber yields a very large increase in the signal-to-noise ratio due to the cosmic radiation and radioactive contamination, as compared with nuclear emulsions. The Wilson cloud chamber has been used to measure the limiting energy of the x-ray radiation spectrum for a powerful pulsed discharge [50]. The energy of the photoelectrons and the recoil electrons produced in the chamber by the x-ray radiation was determined by the track length. As a consequence of the strong curvature of the tracks it is not necessary to measure the total length of the track to determine the energy. It is sufficient to measure the projection on a plane, that is to say, the length of the image of the track on the photoemulsion. In projection on a plane, the projected length of a rather curved track is about $\pi/4$ of the actual length. The spectrum of electrons obtained by irradiation of the sample is compared with the spectrum formed in irradiation of the Wilson cloud chamber by an x-ray tube with a known applied voltage.

References

1. Yu. A. Kucheryaev and D. A. Panov, in: Plasma Diagnostics [in Russian], Gosatomizdat, Moscow, 1963, p. 223.
2. K. D. Sinel'nikov, et al., Zh. Tekh. Fiz., 30:256 (1960) [Sov. Phys. — Tech. Phys., 5(3):236 (1960)].
3. O. A. Lavrent'ev, in: Plasma Diagnostics [in Russian], Gosatomizdat, Moscow, 1963, p. 233.
4. Ya. B. Fainberg, Atomnaya Énergiya, 11:313 (1961) [Sov. Atomic Energy, 11(4):958 (1962)].
5. L. I. Krupnik and N. G. Shchulika, in: Plasma Diagnostics [in Russian], Gosatomizdat, Moscow, 1963, p. 212.
6. O. V. Kozlov, et al., in: Plasma Diagnostics [in Russian], Gosatomizdat, Moscow, 1963, p. 199.
7. Ya. M. Fogel', et al., Zh. Tekh. Fiz., 25:1944 (1955).
8. V. V. Afrosimov, et al., Zh. Tekh. Fiz., 36:89 (1966) [Sov. Phys. — Tech. Phys., 11(1):63 (1966)].
9. Barnett, et al., Atomic and Molecular Collision Cross Sections of Interest in Controlled Thermonuclear Research, U.S. AEC, Oak Ridge, 1964.
10. N. V. Fedorenko, et al., Zh. Eksp. Teor. Fiz., 36:385 (1959) [Sov. Phys. — JETP, 9:267 (1959)].
11. Ya. M. Fogel', et al., Zh. Eksp. Teor. Fiz., 34:579 (1958) [Sov. Phys. — JETP, 7(3):400 (1958)].
12. Ya. M. Fogel' and L. I. Krupnik, Zh. Eksp. Teor. Fiz., 28:589 (1955) [Sov. Phys. — JETP, 1(3):415 (1955)].
13. W. L. Fite, et al., Proc. Roy. Soc., A268:527 (1962).
14. V. M. Chicherov, Zh. Tekh. Fiz., 36:1055 (1966) [Sov. Phys. — Tech. Phys., 11(6):777 (1966)].
15. J. Alexeff, et al., Phys. Rev., 136A:689 (1964).
16. V. B. Leonas, Usp. Fiz. Nauk, 82:287 (1964) [Sov. Phys. — Usp., 7(1):121 (1964)].
17. I. K. Kikoin, et al., in: Plasma Diagnostics [in Russian], Gosatomizdat, Moscow, 1963, p. 193.
18. K. Ehers, Rev. Sci. Instrum., 29:614 (1958).
19. A. V. Chernetskii, O. A. Zinov'ev, and O. V. Kozlov, Instruments and Methods in Plasma Research [in Russian], Atomizdat, Moscow, 1965.

20. L. A. Artsimovich, et al., Atomnaya Énergiya, No. 3, 84 (1956) [Sov. Atomic Energy, 1(1): (1956)].
21. S. Yu. Luk'yanov and I. M. Podgornyi, Atomnaya Énergiya, No. 3, 97 (1956) [Sov. Atomic Energy, 1(1): (1956)].
22. I. M. Podgornyi, N. G. Koval'skii, and V. E. Pal'chikov, Dokl. Akad. Nauk SSSR, 123:825 (1958) [Sov. Phys. — Dokl., 3(6):1208 (1958)].
23. N. G. Koval'skii, I. M. Podgornyi, and M. Stepanenko, Zh. Eksp. Teor. Fiz., 38:1439 (1960) [Sov. Phys. — JETP, 11(5):1040 (1960)].
24. S. Yu. Luk'yanov, I. M. Podgornyi, and S. A. Chuvatin, Zh. Tekh. Fiz., 33:1929 (1961).
25. I. I. Konovalov, et al., in: Plasma Diagnostics [in Russian], Gosatomizdat, Moscow, 1963, p. 154.
26. L. I. Krupnik and N. G. Shulika, in: Plasma Physics and the Problem of Controlled Thermonuclear Fusion, Vol. 3, Naukova Dumka, Kiev, 1965, p. 353.
27. N. G. Koval'skii and V. N. Sumarokov, Zh. Tekh. Fiz., 36:1976 (1966) [Sov. Phys. — Tech. Phys., 11(11):1471 (1967)].
28. V. V. Afrosimov, et al., Nucl. Fusion, Suppl. III, 921 (1962).
29. V. V. Afrosimov, et al., Zh. Tekh. Fiz., 33:205 (1963) [Sov. Phys. — Tech. Phys., 8(2):147 (1963)].
30. V. V. Afrosimov, et al., Zh. Tekh. Fiz., 30:1456 (1960) [Sov. Phys. — Tech. Phys., 5(12):1378 (1961)].
31. Yu. V. Gott and V. G. Tel'kovskii, Zh. Tekh. Fiz., 31:1061 (1961) [Sov. Phys. — Tech. Phys., 6(9):774 (1962)].
32. Yu. V. Gott and V. G. Tel'kovskii, Zh. Tekh. Fiz., 34:2114 (1964) [Sov. Phys. — Tech. Phys., 9(12):1628 (1965)].
33. V. V. Afrosimov, et al., Zh. Tekh. Fiz., 30:1469 (1960) [Sov. Phys. — Tech. Phys., 5(12):1389 (1961)].
34. I. M. Podgornyi and V. N. Sumarokov, Nucl. Energy, C1:236 (1960).
35. L. L. Gorelik and V. I. Sinitsyn, Zh. Tekh. Fiz., 32:1406 (1962) [Sov. Phys. — Tech. Phys., 7(11):1036 (1963)].
36. V. I. Kogan, Zh. Tekh. Fiz., 33:371 (1963) [Sov. Phys. — Tech. Phys., 8(3):273 (1963)].
37. L. L. Gorelik, Zh. Tekh. Fiz., 34:496 (1964) [Sov. Phys. — Tech. Phys., 9(3):386 (1964)].
38. L. L. Gorelik and V. I. Sinitsyn, Zh. Tekh. Fiz., 34:505 (1964) [Sov. Phys. — Tech. Phys., 9(3):393 (1964)].
39. L. L. Gorelik, et al., Dokl. Akad. Nauk SSSR, 147:576 (1962) [Sov. Phys. — Dokl., 7(11):1021 (1963)].
40. Yu. G. Prokhorov, et al., in: Plasma Diagnostics [in Russian], Gosatomizdat, Moscow, 1963, p. 274.
41. P. Raynolds and J. D. Graggs, Philos. Mag., 43:285 (1952).
42. I. M. Podgornyi and S. A. Chuvatin, Dokl. Akad. Nauk SSSR, 117:795 (1957) [Sov. Phys. — Dokl., 2(6):543 (1957)].
43. M. V. Babykin, et al., Zh. Eksp. Teor. Fiz., 47:1597 (1964) [Sov. Phys. — JETP, 20(4):1073 (1965)].
44. U. Grossman-Doerth and J. Junker, Nucl. Fusion, Suppl. III, 107 (1962).

45. B. A. Demidov and S. D. Fanchenko, Zh. Tekh. Fiz., 36:1166 (1966) [Sov. Phys. — Tech. Phys., 11(7):863 (1967)].

46. W. G. Cross, Phys. Rev., 78:185 (1950).

47. I. Alexseff, et al., Phys. Rev. Lett., 10:273 (1963).

48. O. A. Anderson, V. R. Beker, and S. A. Kolgeit, Report at the 3rd International Conference on Gas Discharges, Venice, 1957.

49. V. D. Demichev and Yu. G. Prokhorov, in: Plasma Physics and the Problem of a Controlled Thermonuclear Reaction, Pergamon, New York, 1958, Vol. 4.

50. N. G. Koval'skii, I. M. Podgornyi, and S. Khvashchevskii, Zh. Eksp. Teor. Fiz., 35:940 (1958) [Sov. Phys. — JETP, 8(4):656 (1959)].

Chapter 9

Measurement of the Parameters of
Accelerated Plasmoids

Interest in the diagnostics of fast plasmoids has developed as a result of the introduction of the electrodynamic method of plasma acceleration [1-5] and the appearance of many papers (cf., for example, [5-9]) on the injection of plasma into magnetic traps. Depending on the nature of the problem at hand, it is important to have information on various parameters of fast plasmoids. Work with accelerated plasma leads to certain specific problems and special diagnostic techniques are required. The parameters of the greatest interest are the following: W, the total energy of the plasmoid; P_0, the total momentum of the plasmoid; $\bar{v}$, the mean velocity or, v_{max}, the maximum velocity of the plasmoid (in many cases it is desirable to have the velocity distribution function); N, the total number of accelerated particles and, finally, n, the particle density and T_e, the electron temperature. Under actual conditions of an experiment certain parameters can be determined rather easily while the measurement of others leads to considerable difficulty. However, since all of the parameters listed above are not necessarily independent of each other it is sometimes possible to determine one by the measurement of others.

We shall consider the most popular methods for determining the parameters of a plasmoid.

1. The total energy of a plasmoid is easily determined by means of a calorimeter with partitions. This method has been

described in § 8.4 and does not require further discussion. In determining the time behavior of the energy carried by a plasmoid use is made of the fast method for detection of the heat flux described in the same section.

2. The total momentum of a plasmoid is determined by a ballistic pendulum [3, 8]. The plasmoid strikes the pendulum and transfers its momentum to the latter, $(1 + \gamma)P$, where γ is the reflection coefficient of the plasma from the surface of the pendulum plate. In this case the kinetic energy communicated to the pendulum of mass M is $E = \frac{(1+\gamma)^2 P^2}{2M}$. The mass of the pendulum must be chosen in such a way that its oscillation period will be considerably longer than the duration of the momentum transfer process; on the other hand the sensitivity of the pendulum must not be too small. Usually, M is 10-20 g. In order to eliminate the possibility of twisting of the pendulum due to an asymmetric shock from the plasma on its surface, it is desirable to use a suspension that makes use of two wires. The kinetic energy transferred to the pendulum is defined as Mqh. Here, h is the height to which the bob rises at the maximum deflection from the equilibrium position and q is the acceleration of gravity.

In a spring-loaded pendulum the maximum deflection from the equilibrium position is related to the energy transferred to the pendulum by the expression

$$X = \frac{W}{F},$$

where F is the restoring force of the spring. As a rule, the quantity F is not known and hence the pendulum is calibrated using known values of the momentum.

In the expression for the energy transfer to the pendulum by the plasma there appears a reflection coefficient which, in principle, can vary from zero to unity, although neither one of these limiting cases is realized in practice. Usually the quantity γ is unknown under typical experimental conditions. However, a reasonable value is several tens of per cent. Taking $\gamma = 0.5$ we find that the maximum possible error in the value of the limiting momentum is approximately 30%. If this accuracy is not satisfactory then the reflection coefficient must be determined under the conditions of a given experiment since the numerical value of γ depends both on the type of plate that is used as well as the parameters of the accelerated plasma.

3. The directed velocity of a plasmoid is determined from the time of flight between the detectors which are separated by a known distance along the path of the plasmoid. The detectors can be Langmuir probes operated in the ion saturation current regime, photomultipliers, microwave beams, etc. When the plasmoid moves along the lines of the magnetic field measurements of the time of flight can be made conveniently by recording the change in the magnetic flux expelled from the region occupied by the plasma due to the plasma diamagnetism. The diamagnetic signals are applied to the input of the oscilloscope, these signals originating in coils that surround the moving plasmoid. It is desirable to choose the parameters of the detection apparatus in such a way that the screen of the oscilloscope will show a curve which is not a derivative of the change in magnetic field but the field itself. The problem of integrating the diamagnetic signal has been treated in § 2.3. The use of several magnetic pickup units makes it possible to obtain the velocity of the plasmoid and also to observe the change in shape as a function of distance from the injector. Furthermore, a knowledge of the absolute value of the diamagnetic signal can be used to determine the gas kinetic pressure of the plasma, provided the geometry is known [14].

The most direct and accurate method of determining the directed velocity of particles in a plasmoid is the mass-spectrometer method. A great amount of information is obtained by measuring the deflection of a narrow beam of particles in crossed electric and magnetic fields. The particle beam is formed from the plasmoid through the use of a system of collimators. A spectroscope that makes use of crossed fields has low transmission power, but does provide the possibility of simultaneous determination of the particles in terms of mass and energy.

The velocity of the plasmoid can also be determined by the Doppler shift of the spectral lines emitted by the plasmoid. However, this method of determining the velocity is almost never used since the radiation from a moving plasmoid is usually very small as compared with the radiation of the secondary plasma formed in the apparatus. It is also possible that conditions will exist in the experiment such that the determination of the plasmoid velocity by the Doppler shift is not practical.

4. The total number of charged particles can be estimated if any two of the quantities W, P, and $\overline{v}$ are known. It should be kept in mind that the most frequent error in the determination of N is due to the absence of data on the mass composition of the plasmoid; in the acceleration process not only is the injected gas entrained, but also the products due to evaporation of the walls and electrodes in the injector. Consequently, accurate results can only be obtained through the use of mass-spectrometer methods, which indicate the mass composition of the plasmoid.

5. Any of the methods given earlier can be used to find the density of the plasmoid and the electron temperature. The choice of a method is determined by the experimental conditions and the range of values to be measured.

One of the most reliable and convenient methods for the determination of electron temperature in dense plasmoids ($n \sim 10^{15}$-10^{16} cm^{-3}) that move along the lines of a magnetic field is the simultaneous measurement of the diamagnetic signal and the use of an interferometer which works in the infrared region. The choice of the infrared region is due to the fact that the wavelength is small, so that transmission through a plasma with density $n \sim 10^{16}$ cm^{-3} is feasible, while it is large enough so that small phase shifts corresponding to a density $(2-5) \cdot 10^{12}$ cm^{-3} can be detected. Furthermore, interferometer measurements in the infrared region make it possible to determine the plasma density in the presence of the large amount of neutral gas. For example, at a wavelength $\lambda = 3 \, \mu$ the contribution of ionized hydrogen to the change in the dielectric constant of the plasma (absolute value) exceeds the change due to the neutral gas even when the concentration of the neutral gas exceeds the concentration of ionized gas by three orders of magnitude.

In order to obtain interferograms of a plasmoid the plasmoid is made to move between the mirrors of a Fabry-Perot interferometer. The radiation from a helium—neon laser ($\lambda = 3.39 \, \mu$) enters the interferometer through an iris and, after passing through the interferometer, is detected by an infrared detector (usually indium antimonide operated at liquid-nitrogen temperature).

The background of infrared radiation from the plasma is reduced by means of a diaphragm or a monochromator. Between the interferometer and the laser there is an optical system consisting of a polarizer and a quarter-wave plate. A description of an interferometer operated in conjunction with an infrared laser is given in § 7.2.

We now wish to analyze the data of plasmoid measurements in an electrodynamic injector in the case in which the injector generates two plasmoids, a fast plasmoid and a slow plasmoid (Fig. 9.1).

The dependence of the transparency of the interferometer on the refractive index of the medium, which is required for interpreting the interferometer curve, can be obtained most simply by smoothly filling the vacuum chamber with neutral gas. A typical curve obtained in this way is shown in Fig. 9.2. The interferometer pattern shown in Fig. 9.1 is obtained with an initial phase shift of $\varphi_0 = 45°$. This value of the shift is denoted by the point "a" in Fig. 9.2. As the concentration increases the dielectric constant of the plasma is reduced, the angle φ approaches zero, and the signal from the detector increases. (An increase in the intensity of the detected radiation from the laser corresponds to a displacement of the beam in the downward direction on the oscilloscope.) The peak concentration of the first bunch is achieved at a value φ close to zero. This peak coincides with the peak of the diamagnetic pulse. Then the concentration falls, the angle φ again assumes a value close to the point a, and the signal from the detector becomes smaller. Subsequently in the region of the second diamagnetic pulse there is observed a second burst of concentration which causes the angle φ to go through zero and become positive. In this case the signal from the detector, having reached its peak, then goes through zero. The intensity of the light transmitted through the interferometer approaches the minimum value and the maximum concentration of the second plasmoid is reached at this time. Subsequently the concentration falls and the signal from the detector again goes through zero and then the curve slowly approaches the axis.

A comparison of the experimental data obtained with the interferometer and the diamagnetic unit shows that the electron temperature in the first plasmoid is approximately 20 eV. This

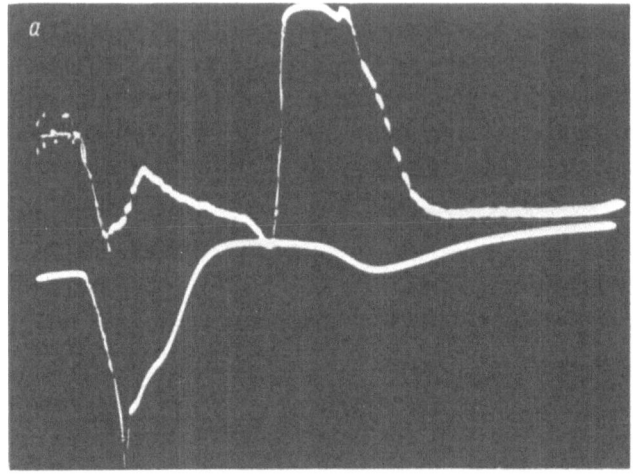

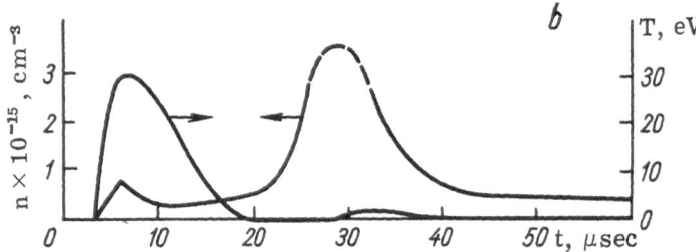

Fig. 9.1. Plasmoid interferogram: phased with the diamagnetic signal
(a) and curve showing the change of temperature and density with time (b).

figure is obtained under the assumption that the gas kinetic pressure of the plasma is determined by the electrons, that is to say, $\kappa T_e \gg \kappa T_i$. The plasma temperature in the second plasmoid is less than a few tenths of an electron volt.

It will be evident that the density and temperature of plasmoids can be obtained by other methods that have been described earlier. However, the advantage of the method just described here lies in its universal nature. For example, the small electron temperature of the second plasmoid from the electrodynamic injector makes it impossible to determine the plasma parameters through the bremsstrahlung because of the appreciable contribution due to the recombination radiation.

The application of other diagnostic methods based on the analysis of the intensity and width of spectral lines is

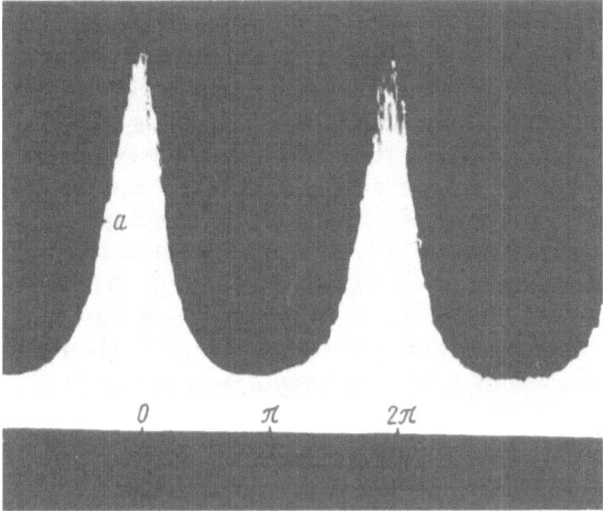

Fig. 9.2. Curve showing the transmission of the interferometer obtained with a smooth variation of the pressure of the neutral gas in the vacuum chamber.

found to be inconvenient for measuring the parameters of the first plasmoid since the plasma is essentially fully ionized and the intensity of the line spectrum is extremely small. In experiments on the interaction of plasmoids with a magnetic field and experiments that study the injection of plasmoids into a trap it is necessary to obtain data on the shape of the plasmoid and the change of the shape as a function of time. Methods based on high speed photography considered in Chapter 4 are not always applicable here since greatest interest attaches to that part of the hydrogen plasmoid which has the highest velocity and which, as is well known, is almost fully ionized. The radiation power of the plasmoid in the visible region of the spectrum is extremely small and photographic methods are not suitable. Under these conditions data on the shape of the plasmoid at various times can be obtained by means of an instrument which has been developed for the investigation of plasma diffusion and which is called the plasmascope [10-13].

A plasma propagating along the lines of force of the magnetic field passes through a gridded electrode beyond which there is another metal grid and a fluorescent screen. On the side of the screen, facing the plasma, there is a transparent layer of alu-

minum of thickness 10^{-5} cm. The plasma electrons are accelerated to an energy of 10^3-10^4 eV in the electric field between the
grids and, passing through the protecting aluminum layer, cause
emission on the screen. Since the accelerating voltage pulse can
be applied at an arbitrary time, the plasmascope can be used to
obtain photographs corresponding to different times. The length of
the exposure is determined by the length of the accelerating voltage pulse and is independent of the fluorescence time of the screen.
A basic shortcoming of the plasmascope in the diagnostics of
plasmoids is the distortion in the density distribution due to the
formation of shock waves in the collision of the plasmoid with the
grid and the formation of cold plasma near the grid.

References

1. L. A. Artsimovich, et al., Zh. Eksp. Teor. Fiz., 33:3 (1957).
2. I. M. Podgornyi, et al., Plasma Physics and the Problem of a Controlled Thermo-
    nuclear Reaction, Pergamon, New York, 1958.
3. G. Marshal, in: Proc. of the 2nd International Conference on the Peaceful Uses
    of Atomic Energy, Geneva, 1958.
4. I. F. Kvartskhava, et al., Zh. Tekh. Fiz., 30:289 (1960).
    Phys., 5(3):266 (1960)].
5. I. M. Podgornyi and V. N. Sumarokov, Nucl. Energy, C1:236 (1960).
6. S. Yu. Luk'yanov and I. M. Podgornyi, Atomnaya Énergiya, 11:336 (1961)
    [Sov. Atomic Energy, 11(4):980 (1962)].
7. J. E. Osher, Phys. Rev. Lett., 10:121 (1963).
8. V. D. Demichev and V. D. Matyukhin, Dokl. Akad. Nauk SSSR, 150:279 (1963)
    [Sov. Phys. – Dokl., 8(5):457 (1963)].
9. G. N. Aretov, et al., Zh. Tekh. Fiz., 34:1191 (1964) [Sov. Phys. – Tech.
    Phys., 9(7):923 (1965)].
10. L. I. Elizarov and A. V. Zharinov, Nucl. Fusion, Suppl. II, 699 (1962).
11. K. D. Sinel'nikov, et al., Zh. Tekh. Fiz., 33:1055 (1963) [Sov. Phys. – Tech.
    Phys., 8(9):786 (1964)].
12. G. M. Batanov, et al., Plasma Diagnostics [in Russian], Gosatomizdat, Moscow,
    1963, p. 263.
13. I. M. Krupnik and N. G. Shulika, in: Plasma Diagnostics [in Russian], Gos-
    atomizdat, Moscow, 1963, p. 256.
14. G. G. Managadze, I. M. Podgornyi, and V. D. Rusanov, Zh. Tekh. Fiz., 37:2199
    (1967) [Sov. Phys. – Tech. Phys., 12(12):1620 (1968)].

Index